A Complete Book on Fish Genetics and Aquatic Environment

A Complete Book on Fish Genetics and Aquatic Environment

Diptanshu Roy

RANDOM

RANDOM PUBLICATIONS

NEW DELHI - 110 002 (INDIA)

A Complete Book on Fish Genetics and Aquatic Environment

ISBN 978-93-51112-81-5

Published in 2014 in India by

RANDOM PUBLICATIONS

4376-A/4B, Gali Murari Lal, Ansari Road
New Delhi-110 002
Phone: +9111-43580356, 23289044
E-mail: randomexports@gmail.com; sales@randompublications.com;
info@randompublications.com

Reprinted 2019

Type Setting by: Friends Media, Delhi-110089
Digitally Pritnted at : Replika Press Pvt. Ltd.

Preface

The successful completion of the human genome project marked the start of a genomics revolution, which has the potential to impact aquaculture and fisheries production and has implications for the future management of fish genetic resources (FiGR). Aquaculture has the potential to compensate for dwindling capture fisheries, but biological and production hurdles must be overcome in order to develop costeffective and sustainable aquaculture. Diseases are major threats to sustainability, and therefore the disease problems should be given high priority. In addition, important production and performance traits (such as growth rate, feed conversion efficiency, body conformation and fillet yield) must be improved in order to make aquaculture more productive and profitable. Genetic enhancement of farmed fish is needed not only to meet the demands of fish production, but also to ensure profitability. The analytical genetic technologies most relevant to aquaculture and capture fisheries include: DNA markers, genome mapping, microarrays, and sequencing. DNA marker technologies are not only the basis for genetic linkage mapping, but also for the analysis of genetic resources, strain differentiation, species differentiation, parentage identification, and preservation of genetic diversity and conservation of genetic integrity. The application of genomics in aquaculture is still at the early stages. For many important species of farmed fish, molecular markers have been developed allowing genetic analysis for FiGR conservation and genetic enhancement of farmed fish. Linkage and physical maps have been developed allowing elucidation of genes responsible for important performance and production traits; genome reagents such as expressed sequence tags have been produced providing material basis for the development of microarray technology. Studies of the genomes of farmed and fished aquatic species have shown both common and unique characteristics that provide both advantages and challenges. In most cases, the genomes of farmed aquatic species are smaller than or comparable to the human genome. Many farmed aquatic species have high fecundity that provides large full-sib and half-sib families, and this greatly facilitates quantitative trait loci (QTL) mapping. However, the large number of farmed aquatic species tends to dilute genomic research efforts.

The genomics revolution and its impacts on aquaculture are expected to contribute to resolving problems such as diseases, environmental impacts, and low profit margins. The major potential applications of genome technologies, primarily in aquaculture but also to some extents in capture fisheries include: markerassisted selection (MAS) for genetic enhancement; environmetal improvements through enhanced productivity as well as the development of novel technologies for environment monitoring, development of effective vaccines and their delivery technologies; monitoring antibiotic resistance; diagnosis for fish diseases and for the safety of aquatic produce; accurate identification of fish stocks for capture fisheries management and for their use as FiGR in aquaculture; conservation of FiGR, including protection of endangered species, in response to fish production strategies and consumer interests; and the development and application of transgenic fish technology including, for example, sterilization technology to address concerns about their possible environmental impacts . A great challenge for aquaculture and capture fisheries is the long-term conservation of FiGR. Genome technologies provide new tools for genetic analysis. Innovative DNA marker technologies have opened a broad avenue for the analysis of genetic diversity based on genotypes. Some aquaculture operations still use wild fish seed. For these and for future fish breeding programs, conservation of wild FiGR is important.

This book has been designed as an introductory text which undertakes a through exploration the concepts and practices which define the subject. The book will prove very useful for students and teachers.

I thank all members of my team who have helped in the preparation of the book. My special thanks go to "Random Publications" who have published the book.

—Diptanshu Roy

Contents

1

Fish Breeding

When Charles Darwin laid down the Theory of the Survival of the Fittest, he was only stating what was so very obvious in nature. The best male of the species will try to impregnate the females, and the females prefer the best male since his DNA increases the chances offspring survival. That is why many animals like the lions etc will have only one dominant male who has successfully defeated all other males in the group. This male alone impregnates the sexually active females. He continues to be the leader until a healthier and stronger male defeat him.

There is something very vital in this theory for the aquarist too. When your fish are ready to breed, it will not do to select just any pair. You need to select the male and female very carefully, unless you are breeding to provide food for other fish. Here are some pointers that will help you choose the right pair:

- Fish that have brighter colours and good markings are the healthiest in a stock. Fish that have good finage, movements and vitality are also healthy.
- The appetite of a fish is often an indicator of its general health and vitality. Only fish that have a good appetite should be selected for breeding.
- Do not use fish that appear stunted or deformed in any way.
- The age of the fish is also important while you consider mating partners. It is better to go for younger fish that are in their prime rather than choosing older fish that have almost reached

the end of the reproductive lives. An exception from this rule can be if you have an older specimen with exceptional code colouration, finage or other qualities that you want to breed on.

- Compatible pairs make good partners. For instance, in Cichlids, fish form pairs only after they have been put together in an aquarium for many months. If the fish are not compatible, one of them may even bully the other to death.
- Unhealthy fish and fish that are not mature enough will not produce healthy fry. You should therefore only choose healthy and mature fish for breeding.
- In case you want to experiment crossing two different strains, please keep in mind that the resulting fry may be unattractive. Never sell crossbreed fish as pure fish. Do not mix impure strains with pure strains.
- Hybrids made by crossing two different species are often sterile. It is possible to mistake the females of different species, because some of them look very similar. Hybrids should be avoided and should never be sold as anything other than hybrids.

Fish Breeding: Condition the Parents

After choosing the parents, you need to condition the fish so that they spawn. One of the most effective ways to do this is to simulate their natural habitats as closely as possible. Making certain changes in the environment will also help you to encourage the fish to spawn.

Providing the right water conditions is one of the first steps. Conditions that are as similar to their natural habitats should be provided. Proper aquarium set-up that provides for enough "security", hiding places and crevices is essential. The water current, lighting and temperature should also be taken into consideration. Some fish mate only when they are in schools. Isolating these fish will discourage spawning.

Providing the right food is the next important thing to do. The mates that have been selected for spawning should be given food that is very high in protein. Live food seems to be a catalyst in this process. You can feed the fish up to three times a day. Of course, remember to clean up the aquarium after each feeding.

Some fish spawn during the rainy season. It can therefore be a good idea to simulate these conditions in the aquarium too. This is actually a very simple procedure. Reduce the water level in the aquarium to half the normal height. Add 5% of the aquarium volume everyday. Add soft water that is slightly cooler that the aquarium water. A drip system or a spray bar will help to add to the notion. During this time, fish should be fed heavily. This process can be repeated till fish show signs of spawning. Most fish respond to the first change in water conditions - like soft water or altered water temperature.

Equipment and Breeding Fish

The first and foremost need for successful spawning of many species is space. Fish often become territorial when their breeding time approaches. Most males are very aggressive during this time, and will chase away any potential competitors.

This becomes especially dangerous for the younger fish, and more so if they are unable to get to a safer place, out of reach of the aggressive male. The primary need of fish that are going to breed is enough space for them to chase, hide and move. There are species that needs very little space and that don't get especially aggressive, such as the very popular Guppy.

Some species of fish get into the spawning mode when they are fed highly nutritious food. Thus it is a good practice to give the breeders some special attention. This kind of special attention aims at bringing the fish to their topmost reproductive state. If you have some extra aquariums, place the two sexes separately in the two aquariums.

A delivery aquarium is advisable when your fish are ready to spawn. This will save the fry from being eaten up as food by the other fish. A tight cover should be placed on the aquarium as some fish species can become very nervous during this stage and show a tendency to jump. The conditions in the delivery aquarium should be as close to those in the original aquarium as possible. This will save the parent fish from any undesirable stress at such a critical stage in the breeding cycle.

A special aquarium for rearing fry is also advisable. This can be the same aquarium as the delivery aquarium if the parents are moved from the aquarium once the eggs are laid. If the species guards their young, they should not be removed until they have stopped rearing the fry and started to prepare for a new spawning. The needs and requirements of fry are very different from bigger fish. They need special kinds of food, at least till they grow a bit bigger. Filtration too has to be varied in order to accommodate fry.

As the fry start developing you can perform a 10% water change the first few days, and than raise this up to a 40% water change every day to keep the water clean. Clean water helps keep your fry healthy and fast growing. Optimal survival of the fry is ensured if you segregate the fry depending on their size. The fastest growing fry are almost always the males. The bigger fry will eat the smaller ones if they are not separated.

Some people use a breeding trap as the place where the female livebearer can drop her fry. The fry are thus out of reach as they will be dropped into a separate plastic compartment attached to the rim of the aquarium. If the breeding aquarium is too small to accommodate the female, the chances of her miscarrying while struggling to escape are very high.

The spawning aquariums for different kinds of fish should be set up according to their particular characteristics. Egg scatterers will eat their own eggs. So, care should be taken that their eggs fall away from the reach of the parents. For small species of this kind of fish, a 5-10 gallon aquarium would be the best. The parents should be removed after spawning. If the fish lays non-adhesive eggs, it is better to put nylon netting just above the aquarium floor. The eggs will then fall out of the reach of the parents. In case the eggs are adhesive, then they will be laid among the leaves of plants.

Egg depositors should get aquariums decorated in accordance with their egg depositing behaviour. If they are the types that care for their young, the parents can stay in the aquarium after laying the eggs. It is advisable to provide these fish with broad-leafed plants, flat stones, upturned flowerpots, coconut shells, or rocky caves. If your fish species inhabit rocky areas in the wild, they will appreciate caves or flat stones as breeding sites, and so on. Putting in plants and caves will give fish a sense of security even if they are not plant spawners.

Mouth brooders can be left in the main aquarium even when they are spawning because the eggs as well as the young are well protected by the parent. Brooding does however bring out aggression and brooding fish can cause harm to any other fish in the aquarium. For this reason alone, it may be better to put these fish into a new aquarium while they are breeding.

Setting up a new aquarium for breeding purposes is fairly easy. You do not have to go through all the steps and wait for the Nitrogen cycle, plant your aquarium etc since you can use water and bacteria from your established aquarium. An additional aquarium for breeding purpose need not be very big. Usually an aquarium with a capacity of 2-6 gallons would suffice for most fish.

The best way to set up your aquarium easily is to take water from your existing aquarium and use this to start your new aquarium. Since the fish are being transported from the old aquarium, they will already be acclimatised to this water. Using a foam filter for a few weeks in your old aquarium will help you collect a good quantity of beneficial bacteria that can be easily moved to the new aquarium. Decorations and plants can be kept to a minimum for most species. Changing water in this aquarium regularly is a must.

The best way to clean out the material wastes in this aquarium is to siphon it out using a flexible pipe. Keep in mind that it is easy to siphon out some fry also along with the waste. So, it is best to siphon out the waste into a clean bucket. This will help you to save any nosey fry.

Failed Breedings

Reproductive failure is one of the most frustrating experiences for the aquarist, especially a beginner. These failures can occur at any stage of the reproductive cycle. The first major stumbling block for the beginner aquarist is failure to get the fish to spawn. For success, some species need the perfect set up.

A deep knowledge of the species, the right food, the right temperature for the species you select in addition to any necessary spawning triggers have to be used efficiently. For the beginner it is also important to choose the right species to breed as some species are much easier than others to breed. A separate spawning aquarium with or without plants according to the needs of your species is the first step in the right direction. Novice breeders also need to be very careful with the water chemistry. You cannot just put your selected pair in a tub of water and expect them to get on with their work. You have to get a fully working testing kit and learn how to use it. Higher quality in the breeding stock directly influences the viability of the spawn. The quality of the offspring will be affect by the kind of water in which it was bred.

Sexing the fish correctly is the next big step. In some species, differentiating between the two sexes is very difficult indeed. So, this is not to be taken lightly.

Do not keep checking on your fish every few minutes. This will make them stressed. Stressed fish will not spawn. Fish that are too young or too old must be avoided at all costs. Left to itself in nature, the female fish will select the male fish that is in its prime. Males that are too young or too old will not have viable sperms, and the eggs that are laid as a result may not get fertilized.

Once the fish has spawned, problems may arise at the time of hatching too. The biggest impediment to hatching is water quality. Bad water quality is not conducive to spawning, and even if the fish do spawn, the eggs will not develop properly.

As a result the eggs may catch fungus and die. If the eggs are removed to a hatching aquarium, care must be taken that they are not exposed to the atmosphere for long. The eggs dry up when they come into contact with air.

The water in the hatching aquarium must be as similar to that in the original aquarium as possible. Sudden changes in temperature will halt the development of the egg. Hard water will cause the shells of the eggs to harden, and in very hard water the shells can become too hard and make hatching difficult or impossible. If the water is too soft, the eggs may instead collapse.

Sterility is also a common cause for reproductive failure. In some species, it is seen that the absence of any male will cause two females to pair off. Obviously, no eggs will be fertile after such a spawning. Pairing off a female with a male that is too old or young will also cause the eggs to be underdeveloped. Again, hatching of such eggs will never take place.

Disease and vitamin deficiency are some other factors that play a decisive role in the ability of a fish to spawn. Over-feeding the fish and making them obese will rob some male species of their virility.

There are some varieties that are highly suited for the novice to breed. As stated earlier, livebearers are among the easiest to breed and will most likely spawn spontaneously in your aquarium without you even trying to coax them.

Of these, the Guppy, the Molly, the Platy and the Swordtail are the easiest to spawn. Mollies and Guppies are very similar in their breeding patterns. However, newborn Molly fry are susceptible to protozoan infection.

Sometimes, the Molly may also give birth to unformed fry that appear almost at the same time as the swimming fry. They appear as white eggs and must be disposed off without delay. Among egg-layers, the novice may want to try Steel Blue Killifish, some Dwarf Cichlids like Kribs, the Tiger Barbs, various Danios or the White Cloud Mountain Minnow.

Obviously, breeding fish is a reasonably challenging task, but it can be achieved quite easily, if the right steps are followed. Remember that as in humans the female carrying the eggs as well as the eggs themselves must be in the best possible surroundings. Maintaining the right condition inside the aquarium will help the dwellers of the aquarium to go about their work in a healthy manner.

Feeding Fry

Infusorians

Infusorians is a suitable food for very small fish fry. You can easily grow your own Infusoria cultures at home. This way you will always have cheap and nutritious food ready for your fry. Infusorians are minuscule single cell organisms that live in water. To cultivate your own Infusoria you will only need tap water, some bruised vegetables and a few clean containers such as jars or similar.

Boil the tap water and let it cool down to room temperature. Fill the jars with the water and add some bruised vegetables to each jar. Bruised lettuce leaves or banana skins are recommended, since they are known to produce a lot of Infusoria. One banana skin or 3-4 lettuce leaves is a good amount for each jar. Do not put any lids on the jars. Place your jars in a warm place where they will receive moderate light.

The water in the jars will soon turn cloudy and begin to smell a bit. As the Infusorians develop, the smell will become increasingly sweet and the water will clear up. It will usually take 3-4 days for Infusoria to develop. When the water looks clear, you can start feeding your fry with the Infusoria.

Use a siphon or a spoon to remove the Infusoria from the top of the jar and place it in your fry aquarium. Make sure not to bring any plant material from the Infusoria jars over to the fry rearing tank, since this can pollute the water. Start developing new Infusoria culture every 3-4 day to make sure you always have enough food for your fry until they are big enough to eat larger food.

Brineshrimps (Artemia Salina)

Most fish fry is large enough to eat brine shrimp as their first food, especially if you give them newly hatched brine shrimp. Very small fry should instead be fed Infusoria, and brine shrimp can be introduced as a second food.

If you feed your brine shrimp suitable food, they will grow large enough to be used as food for adult fish as well. Growing large brine shrimps is an inexpensive way of providing your fish with much appreciated live food, and is especially recommended when you wish to induce spawning.

Just like Infusoria, brine shrimp is easy to cultivate at home. You can buy a brine shrimp hatchery from your fish store or fix one yourself. Brine shrimp will even hatch in bottles! Some brine shrimp hatcheries will require and air stone connected to an air operated pump. Other brine shrimp hatcheries can be attached to the inside or outside of the aquarium.

If you want to fix your own brine shrimp hatchery in a bottle you can use a clean milk bottle or similar. Boil some water, three-quarters of a pint is enough, and let it cool down before you use it to fill the bottle. Add one teaspoon of aquarium salt to the water and put the bottle in a warm room.

It is possible to use ordinary cooking salt instead of aquarium salt, but the result will usually not be as good as with aquarium salt. Insert an air stone in the bottle to provide oxygen. Place a pinch of brine shrimp eggs in the water and wait for them to hatch. The water temperature must never go below 15 degrees Celsius, but this is usually not a problem in a temperate room. Brine shrimp eggs will typically hatch after 48 hours or even earlier. By setting up new bottles you can make sure that you always have suitable size brine shrimp for your fry.

Brine shrimp eggs can be bought in most pet shops and fish stores. If you use eggs with shells you must turn off the air stone in the hatchery to make the shells sink to the bottom. The hatched brine shrimp will stay a few inches above the bottom and can be easily removed. It is important to separate the brine shrimp from the shells, since shells might get stuck inside the fish fry if they consume them. You can also buy shell-free brine shrimp eggs.

Water Fleas (Daphnia)

Water fleas can be hard to disinfect properly, so if you want to feed your fry daphnia you must obtain the daphnia from a source that you can trust and that you know have no sanitary problems. The same is true if you plan to raise your own daphnia for the fry; you must make sure that you begin with a disease free starter culture.

Once your have obtained a proper starter culture of daphnia, the rest is easy. Fill a plastic container with 5 gallons of tap water and let the water stand for roughly 48 hours in room temperature. Obtain a handful of manure from a stable.

You will also need a nylon bag. You can produce one yourself from a pair of nylon stay-ups or use the type of nylon bag that comes with washing machine tablets. Place the manure in the nylon bag and submerge the nylon bag in the water. After roughly 7-10 days the water will have turned cloudy, which means that the water is now filled with micro organisms for the daphnia to feed on.

Add the daphnia to the water and wait. The culture will take care of itself for the few weeks needed for the daphnia to grow large and abundant enough to be suitable as fish food. The easiest way of removing daphnia is to use a fine meshed net.

Whiteworms

Whiteworms are a much appreciated fry food that can be cultured at home. You will need to obtain some moist and nutritious garden soil for the whiteworms to grow in. Fill 75% of a shallow box with garden soil. If necessary, water the soil until it is quite damp.

It is important that the soil never is allowed to dry out, but waterlogged soil is just as bad. Add the whiteworm culture to the soil together with a few small pieces of moist white bread. Cover the box with a lid. The box or the lid must have some air holes to allow for ventilation. Put the box in a dark place where the temperature is around 16-18 degrees Celsius. Bring new bread to the box every 3-4 day and make sure that the soil is constantly moist. If you find any uneaten food you should remove it before it turns bad.

When its time for your to collect whiteworms for your fish, you just spoon out some soil from underneath the bread. You will naturally find the largest congregations of whiteworm right beneath the food pieces. Pour the soil with the whiteworms into a bowl or dish filled with water. When the worms have become separated from the soil you can easily collect them and drop them into the aquarium.

Fish Spawning Methods

A majority of the world's fish species are egg layers, but there are also a lot of species that have developed into livebearers. The egg laying species will typically use some form of external fertilization of the eggs, while the livebearers perform internal fertilization. The level and type of parental care will also vary considerably between different fish species. Some are extremely devoted parents while others leave their offspring immediately, or even worse – try to eat them! The parental care can also vary a lot between different individuals within the same species, and from batch to batch.

Livebearers

Livebearing fish species give birth to free swimming fry. A lot of the commonly kept aquarium species are livebearers, such as the Guppy. The eggs are fertilized by the male while they are still inside the body of the female fish, so called internal fertilization. In livebearing species, the male's anal fin has developed into a reproductive organ. This type of modified anal fin is called a gonopodium. After the fertilization, the eggs develop into fry inside the female fish. When the fry is born, they look like miniatures of adult fish, but sometimes without the striking colours of their parents. Many species develop striking colours as they grow larger and mature. In some livebearing species the female fish can store spermatozoa (semen) from one single mating and use it to fertilize several batches in a row. This is why female livebearers sometimes give birth in aquariums where they are kept without any males.

Egg Scatterers

Egg scattering species are often living in fast flowing waters in the wild. The eggs are released by the female and fertilized, and then rapidly swept away by the current. This is why egg scattering species will usually eat their own offspring if they spawn in aquariums. If you are planning to breed an egg scatterer, e.g.a Zebra Danios, you must protect the eggs from the parents. You can cover the aquarium floor with marble and keep the water shallow. The eggs will sink and since the water is shallow they will rapidly reach the bottom, where they will lay hidden among the marbles. Larger marbles will also provide hiding places for the fry when they emerge. Some breeders place a net in the aquarium, to prevent the adult fish from reaching the bottom. The mesh must be large enough for the eggs to fall through.

Egg Depositers

There are two types of egg depositers: open water spawners and secretive spawners. Open water spawners will not hide their eggs. Open water spawners typically deposit the eggs on a flat rock, on the aquarium floor or on the leaves of broad leaved aquatic plants. Some open water spawners will dig a crater in the substrate and place their eggs inside. Secretive egg depositers wants to hide their eggs, and will choose a cave, a crevice or similar as spawning site. In the aquarium they will usually appreciate turned over terracotta flowerpot. Both open water spawners and secretive spawners clean the site before they deposit their eggs, and they will usually also defend the eggs and care for them by cleaning them. Before the fry emerge, many egg depositers will move the eggs around to other pre-cleaned sites. Egg depositers are also known to guard their fry and the fry is often seen swimming around with their parents.

Nest Builders

The nest building group is actually a sub-category to the egg depositers. Nest builders create nests of bubbles and deposit their eggs inside. The nests are formed by the male fish by blowing saliva bubbles, and a bubble nest will often contain plant fragments. When the female has released the eggs they are carefully gathered by the male and placed inside the bubble nest. In some species, the female will gather the eggs. In some species the eggs will float up into the nest without help from the parents.

Mouth Brooders

Mouth brooders are also a kind of egg depositers, but the offspring is guarded inside the mouth of a parent. The eggs are often deposited in a shallow crater where they are fertilized by the male before being picked up by the male or the female fish. In some species the eggs are fertilized by the male when they have already been picked up by the female.

Some mouth brooding species will carry only the eggs inside the mouth and spit out the emerging fry, while others will care for the eggs on the spawning site and pick up the fry with their mouth when they emerge.

Some species will carry both egg and fry. Even after being released, the fry might flee into the mouth of their parent if they are scared by something or sleep inside the mouth during the night.

Egg Buriers

The egg burying species have developed in areas subjected to seasonal draughts. During the rainy season, the eggs will be deposited deep down in the mud. The eggs will then stay buried during the draught. When a new rainy season begins, the water will trigger the hatching of the eggs and fry will emerge. The egg burying species will not guard their offspring, since the parents are usually dead long before the rainy season begins.

If you want to breed egg burying species in your aquarium you must simulate a dry season and a subsequent rainy season. The adult fish can be given a container filled with peat moss to deposit their eggs inside instead of placing them in mud. The container can be removed from the water and subjected to a several month long dry season in a cupboard. Once you add water to the peat moss again, the eggs will hatch and you can start feeding the fry.

Koi Breeding

Koi breeding is a fascinating but can sometimes be quite hard since Koi fish grows quite large and because you will have to have good parents to produce quality Koi offspring.. This article is meant to explain some essential aspects in the breeding of Koi fish. All the described facts are very important so please don't mind the order in which they are written.

The first consideration when breeding Koi is to get suitable fish. High quality Koi will produce the highest quality offspring and you will have to choose the parents after which kind of offspring you want in regards to colouration and pattern. Only a very low percentage of the fry in the large batches will become high quality Koi regardless of the quality of the parents.

High quality parents will however produce a higher percentage quality fry. It is recommendable to try and keep the bloodlines of you Koi type clean to produce high quality Koi fish.

Another aspect that should be taken into consideration while choosing the indented parents is the age of the fish. As this effects the quality of the egg shells. Young Koi females produce eggs with very thin shells which might not survive; older fish on the other hand can produce eggs with to hard shells that the sperm can't penetrate them. The best age of a Koi that is used for breeding is 4-5 years although successful breeding might be possible with fish that are up to 15 years old.

The chosen couple should be separated from other fish of the other sex for at least a month during which time they are kept well feed to get them into spawning condition. Koi breeding is often best done with pairs of fish but you can use 2 males to a female while breeding Koi. You should however never use more than two males since they may become too aggressive and hurt the female if more than two males are used.

The best time of the year to breed Koi if it is done in out doors is different depending on where you live in the world it is however recommendable to use a time of the year when the temperature difference between different times of the day is as smallest. Koi breeding are most likely to be successfully during a full moon since this helps trigger the fish to spawn.

Koi Breeding Tank

You will have to decorate the aquarium so that there are suitable breeding areas present for the Koi fish. The materials selected to be used for spawning should be totally submerged and safe to the environment. There are many synthetic Koi breeding materials available in the trade but most professional breeders still prefer the natural ones and think the produce better results. An example of a good natural breeding material is Water hyacinths which are a plant that can be great for building suitable spawning places with. Water hyacints can carry parasites and will have to treated with special substances like Dipterex to eliminate all possible parasites before being used in a Koi breeding setup. Water hyacinths might also attract dragon flies. While these insects are harmless, their larvae can attack and kill the Koi eggs and fry. It is recommendable to use nettings on top of your breeding pond if you breed your Koi outdoors to protect them from predators. It might also be beneficial since they sometimes like to jump and netting prevents them from jumping out of the pond. A benefit that makes the use of nettings on your pond a beneficial choice regardless of whether you keep and breed your Koi indoors or outdoors.

The pH level in the tank is not that important in regards to getting the Koi to spawn but will effect the quality of the fry. A neutral water is to be preferred as acidic water causes the fish to get good red colour but weak black colour and an alkaline water will cause good black colour put weak red colour.

The good Koi breeding pond should have a round shape and around 1,000 gallons / 3800 L of fresh and chlore-free water. The water will if Chlorated have to be treated with Clorox or another equivalent to remove the chlore before being used to breed koi. The water should be kept standing for three days before adding fish. The water should be aerated using some strong water pumps. A strong current or a splashing waterfall adds to the aeration but can have a damaging effect on the spawning.

The male will try to push the female against the pond walls and other objects to force the eggs out of here during the spawning. He fertilizes the eggs once they are released. If two males are used they often push the female from each side to get her to release her eggs. One spawning results in many thousand sticky eggs that adhere to anything they touch. It is good to keep a watchful eye on the spawning since the male sometimes injures the female in which case you will need rescuing. The parents will eat their own egg and fry so the It is recommendable to move the male and the female to different tanks and to give the female a chance to rest after the breeding. The female will still smell like spawning and males will keep trying to spawn with her if you are kept in tanks containing males. This can often lead to the males hurting the female. The spawning smell will have vanished after about 24 hours. The male will also have to recuperate since he otherwise might become infected with a diseases and fungus

The fry will survive on their yolk sac their first day but will after that need to be feed 4-5 times a day for the following 3 months. Suitable food includes infusoria and algae (supplement food only) and later newly hatched brine shrimp.

As the fry grow you can start switching them over to regular Koi pellets. Make sure that you keep the water in good condition as the frequent feedings put a large strain on the water.

It is often almost impossible to raise all the fry in a spawning and it is recommendable to only raise half the fry or to move half of them to another pond since the breeding pond as described above otherwise becomes too crowded. It is also recommend to sort the fish after quality as they grow and only keep the fry of higher quality.

Raising Fry

Once the eggs are laid, you can remove the parents in none fry guarding species. This is especially important if the fish are the types that eat their young. In case of fish that care for their brood, it is recommended to keep the parents with the fry.

Taking care of your fry requires some special attention. Some young ones, like the Goldfish, will not eat for the first 40-48 hours of their life, while other species need to be fed immediately. How do you know when to do what? And what are the best conditions for keeping the fry alive. These are some of the many things that you need to keep in mind when the fish start spawning.

In some livebearers, the female may release only a few of the babies at the beginning. She may take a few hours or even a day to release the entire batch of fry; and some of these fry will likely be stillborn or even eggs.

Even after this, a female can be full of fertilized eggs that will soon begin to develop into fish. Livebearers can release several more batches of fry even if there are no males around, due to their ability to save sperm.

The fry may take up to a few weeks to develop. Do not expect your fry to come out ready for a swim and a feed. The larvae that hatch have almost nothing in common with their parents. They will be enclosed in a big yellow yolk sac and will not be able to swim at all. The larvae will eat up the egg sac until the yolk is fully gone. Once this is over, the fry begin to get hungry and look for food.

Infusoria is a good food for tiny fry. Green water or egg yolk can also be fed to the fry. Powdered flake food is commonly fed to fish fry in aquariums, but powdered flake food is not the best fry food. Powdered flake food is however better than nothing if it is all you got when the fry arrives. Take a plastic bag and put in 1-2 tablespoons of flake food. The bag should not have too much air in it. Then squeeze the bag in your hand and mash the flake as well as you can. The more powdered it is, the better.

Putting in only a small amount of food for the fry is also a bit difficult. The best way to add in small amounts of powdered food is by using a toothpick. Wet the end of the pick in water, and wipe it with your finger so that it is only a little wet. The wet toothpick should now be dipped into the powdered flakes. Touch this tip in the water just above the fry. You will need to feed them several times a day.

This powder is quite fine and is acceptable for many fry to start off with. But some fry find even this too big to eat. For instance, the newly hatched Angelfish cannot eat this powder. It needs to eat infusoria or baby brine shrimp for the first few days.

Baby brine shrimp is a very important food for most fish fry. Most livebearers will eat lots of baby brine shrimp, and this definitely makes them grow faster and healthier even if livebearers are among the fry that can be raised on powdered flake food alone. The eggs of Brine Shrimp are carefully processed and collected and are sold in many pet stores. The aquarist can purchase these eggs, hatch them and use them to feed fry. One advantage in hatching the eggs at home is that these can then be fed special additives that will directly be transferred to your fry. Some fry are too small to be able to eat newly hatched brine shrimp and such fry will need infusoria the first few days before they can start eating brine shrimp.

After they start feeding, the fry usually grow pretty fast. Feeding should be done several times a day for optimal growth. Aquariums that are well lit, well planted and nicely filtered provide the best conditions for rearing fry. Having a fixed feeding schedule also helps the fry grow faster. Frequent water changes are a MUST if you want healthy fry. Remember that the filtration in an aquarium that contains small fry will be less efficient than the filtration in a proper aquarium as the circulation has to be kept lower. The number of feedings is also higher and the wastage will also be more. Thus frequent water changes; stable water chemistry and sufficient warmth contribute greatly towards the well being of your fry.

Spawning Methods

Sexually, fish can be classified into two groups: the sexually dimorphic and the sexually isomorphic species. The sexually dimorphic species show obvious primary and secondary characteristics of their gender. The males of this species are often larger and more colourful, and they usually have a much more elaborate finage.

The sexually isomorphic species show hardly any apparent differences in the two sexes. It is therefore very difficult to distinguish between the two sexes in this group. Often, the only deciding factor could be the shape of the genital papilla. This becomes visible at the time of spawning. In some of these species, the female have a more rounded belly or is slightly larger than the males. There are some species that show no visible difference at all between the two sexes.

External fertilization followed by egg laying is the way most fish in aquariums will reproduce. Egg layers are classified into five groups:

Egg Scatterers

These species, as the name indicates, scatter their eggs on any surface. The eggs may be adhesive or non adhesive. After laying eggs, the parents do not look after them, and may even eat their own eggs. These are mostly schooling fish, and they often spawn in groups. Their eggs hatch fairly quickly.

Egg Depositors

These fish deposit their eggs somewhere on the substrate. They lay fewer eggs than egg scatterers. Some species within this group show advanced care for the eggs and fry. They will deposit eggs in an enclosure, cave, a pit or similar, and protect and clean the eggs. Some species of Cichlids come under this category. Not all egg depositing species care for their young. Some just deposit their eggs and then abandon them. Egg depositing species generally don't eat their own eggs. Their eggs are usually larger than those of the scatterers.

Mouth Brooders

There are two basic types of mouth brooders. In some species, the female fish carries the eggs and/or larvae in its mouth. The female in this group receives a mouthful of sperm from the male, and the fertilization takes place in the mouth of the female. In the second type of mouthbreeders, the females lay the eggs on a substrate and guard these eggs till they hatch. Once the eggs hatch, she picks up the fry and carries them in her moth till the fry have become a little bigger. There are also paternal mouthbrooders where the offspring is carried by the male fish.

Nest Builders

These fish build a kind of nest that is made of plant debris and saliva. They may also dig a pit in the substrate to lay eggs. They take care of their young ones.

Livebearers

Livebearers are fish that do not lay eggs. The fry come already hatched out of the mother fish. Livebearers are able to give birth to young ones because of internal fertilization which takes place inside the female. Sometimes contact between the cloacal apertures of the male and female fish leads to fertilization. In some species, the eggs produced by the livebearer will be heavily yoked. In other species, the young are in direct contact with the mother and are fed and nourished by her. In this case, the mother carrying the young ones needs more nutrition and overall care. Fry of livebearing fish usually takes about 30-32 days to develop. Many factors will affect this development period, e.g. the water temperature, the female's age and her level of nutrition. The good thing about livebearer females is that they are immediately ready to breed again after they drop their young ones. The competition to get to these females is intense among the males.

For beginners livebearers are the easiest fish to breed. If given enough space, the livebearers will do most of the hard work. Once the fry become noticeable and have started swimming, they need to be removed from the aquarium unless the aquarium is densely planted. Failure to do so will surely land a large percentage of them into the stomachs of bigger fish. Many people breed livebearers to provide food for predatory fish. Some livebearers will cross breed with other fish and to avoid hybridisation these species should not be kept together.

Tips for Effective Fish Breeding

Anything that you have placed in the aquarium traps debris. As soon as the fry come out, remove any breeding traps that have been in the aquarium before.

You can also add one or a couple of apple snails into the aquarium. Apple snails are very laid back and do not attack any fish. These snails will eat debris, and will also eat any dead fry. Apple snails will not eat live fry. Apple snails also produce a good supply of infusorians and can even provide the frys first supply of food.

Keeping a detailed log of the entire process is a very good idea. This will help you when you need to repeat the process. You should keep accounts of the species name, the detailed water chemistry, filtration and aeration methods, approximate ages of the parents, when the female was added, the date on which the fry was released/ hatched, the frys first food and the body size of the fry till it reaches about three months of age.

Sometimes, even in spite of doing just about everything to ensure success, you will find that your fish just does not breed. All conditions are as they should be, but still there is no success. In this case, it is best to use a target fish to jump-start the process.

This works best in territorial fish. A target fish is a perceived threat to the territorial fish. The male fish sees the new fish as a possible encroacher in his territory and he will then pair up with the female fish to isolate the target fish.

The important thing here is to ensure that your target fish is in reality not a threat to the existing fish or vice versa. For instance, when trying to breed small cichlids, using some zebra danios as target fish will not pose any threat to either fish.

It should be noted that some species of fish will kill any kind of target fish, and such situations should be avoided. If this is the case, it will also suffice if you place the target fish in a different aquarium alongside the mating aquarium so that the male can see but not touch.

Sometimes, fish introduced into a new aquarium are too nervous to come out into the open. They will remain hidden for days and will refuse to settle down. Fish that are stressed in this way will never pair up and breed. In such cases, you can use a dither fish to calm the other fish.

A dither fish is an easygoing, harmless fish. A hyperactive and aggressive fish is definitely not the right dither fish. Calm and peaceful midwater fish makes the best dither fish.

Once the nervous fish see the dither fish swimming about happily without being consumed by predators, they too will settle down and come out of their hiding places. This is just to reassure the nervous fish that nothing will harm them when they come out into the open.

2

Elective Breeding

Selective breeding proved to be the most effective process for genetic improvement of plants and animals. Achievements in increased yields of paddy, wheat, milk, meat etc. during recent decades are due in large to selection. In aquaculture, the potential has not been fully exploited through selection. However, the selection work on Atlantic salmon in Norway, th Channel catfish in the USA and the Nile tilapia in the Philippines is enough testimony to advocate that it is quite possible to improve many economic traits like, survival, growth, resistance to disease, meat quality etc. through selection. In fact, many hatchery breeding programs result in lower quality fish through inbreeding depression.

In most of the developing countries, particularly in India and other South-east Asian countries, a substantial number of people depend on fishing or fish culture for their living. Consequently, there is over-exploitation due to heavy fishing both in marine and riverine areas reducing the size of populations in the absence of proper replenishment of the depleted stocks.

In aquaculture ever since the success of induced breeding of the Indian major carps through hypophysation, the technique has been gradually popularised throughout the Indian subcontinent as well as neighbouring countries such as Pakistan, Bangladesh, Burma, Nepal etc. Consequently, many hatcheries have developed in almost all these countries. However, in over four decades, attention has been paid only to produce fish seed to meet a pre-targeted quantity with very little regard to quality.

A survey of carp hatcheries in the state of Karnatak in South India was made (Eknath and Doyle 1990) because of the growing concern that hatchery stocks may be inbred, as suggested by rapid deterioration of their performance.

Eknath and Doyle (1990) found that the rate of the accumulation of inbreeding (F) was particularly high for the three most desirable carp species in the country, catla, rohu and mrigal.

The most important reasons for this appeared to be that all the hatcheries function as isolated genetically closed units, raising their own stocks of brood fish.

Moreover, the individuals for replenishment of brood stock in the hatchery are those which were retained after the hatchery had reached its seed production target for a given year (Eknath and Doyle 1990).

It was also reported that small hatcheries appear to rapidly inbreed their stocks.

The tendency to breed slow growing and later maturing individuals was also observed in some of the hatcheries, which is undersirable.

The survey Basavaraju *et al.* (1997) conducted particularly for catla in the three major seed producing hatcheries in the state of Karnataka, namely Tunga Bhadra Dam (TBD), Bhadra Reservoir Project (BRP) and Kabiri Reservoir Project (KRP), also indicated that the practices followed by these hatcheries are likely to result in both inbreeding and negative selection.

Maheshwari and Biradar (1998) reported decline in fertility of hatchery bred *Labeo rohita* due to inbreeding.

A study was conducted at Powarkheda, Freshwater Fish Farm of the Central Institute of Fisheries Education (ICAR), India where induced breeding of *L. rohita* was carried out between 1982 and 1992 to assess fertility.

It has been reported that breeding of related individuals up to the sixth generation led to 71% decline in fertility level as compared to base population.

It was also observed that the range, standard deviation and coefficient of variation were highest for fecundity in the first generation. This indicates, according to the authors, the scope for improvement through selection.

Selection Work on Indian Major Carps in India

Currently two projects are in operation on carp selection. One on *Labeo rohita* as already mentioned, the other on *Catla catla* at the University of Agricultural Sciences, Bangalore, Karnataka State.

Selective Breeding of Rohu

Founder populations of *L. rohita* have been collected from five different rivers in North India, namely, Rivers Ganga, Gomati, Yamuna, Sutlej and Brahmaputra, which are the native habitat of rohu. Rohu stock from the fish farm of CIFA has also been added to the gene pool. Thus the base population for the selection project consisted of rohu, from six sources, to broaden the gene pool.

All these stocks were collected from fry to fingerling stages and reared separately in earthen ponds for a period of one to two months before they were physically marked and stocked in communal ponds for raising to brood fish.

Marking was done using dye, fin clipping and a combination of both. Ganga rohu were dye injected with M-procian blue at the base of the right pectoral fin and Yamuna rohu at the base of the left pectoral fin.

In the case of Gomati and farm rohu the right and left pelvic fins were clipped respectively, while the Sutlej rohu were marked with the blue dye, inject at the base of the left pectoral fin by clipping the right pelvic fin. The reverse has been done to mark the Brahmaputra rohu, i.e. by injecting blue at the base of the right pectoral and clipping the left pelvic fin (Gupta, 1994).

During 1993, the first groups of full sibs were produced from the local and Ganga stocks as these two stocks were matured. Two or three males were mated with one female (paternal half sibs).

However, during the year 1994, full-sib groups (families) were produced from all the six stocks of rohu: two or three females were mated to a male, producing maternal half-sibs.

Rearing full-sib groups to tagable size fingerlings has been done in earthen nursery ponds of 100 m^2 for four to five months (Saha *et al.*, 1997). After the full-sib attained suitable tagging size (8–10 g) the fish were tagged by using PassiveIntegrated Transponder (PIT) tags, before they are stocked in ponds for communal rearing under both mono and polyculture systems. (Mahapatra *et al.*, 1998).

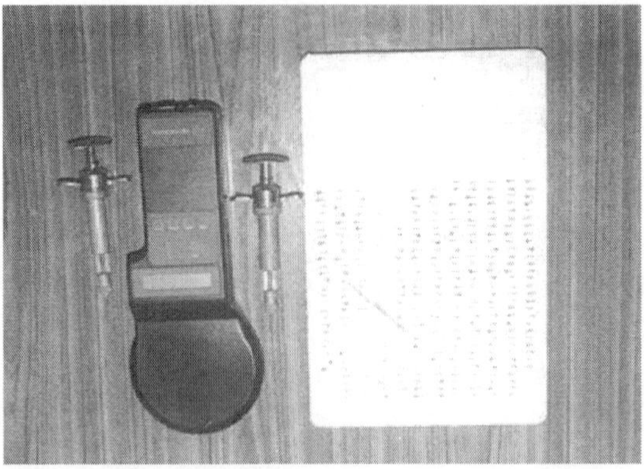

Figure : *Passiveintegrated transponder (PIT) tags, tag implanter and tag reader*

After a period of one year, data pertaining to growth (length and weight) was collected and analysed to evaluate the individual breeding value and to rank the fish. The data so far collected indicated that growth and survival of the wild rohu stocks were equal to or better than the farmed stock.

This indicated that the present procedures followed by hatcheries for seed production are improper. A substantial additive genetic variation has been observed for growth rate and survival of rohu, both under mono and polyculture systems, thus indicating that growth rate and survival of rohu can be improved through selective breeding.

Competition for food may prohibit rohu from exhibiting its growth potential in polyculture. Proper feeding and pond management is essential while testing fish in a breeding programme.

The study so far has also indicated that stocks and full-sib groups within stocks ranked consistently high for growth performance in mono and polyculture, which suggests that there is no need for the development of special rohu lines for different systems. However, further studies are envisaged to test genotype by environment interaction, i.e. under intensive/extensive production systems and in different climatic zones.

Selective Breeding of Catla

The main objectives of this project are to determine the extent of the problem of inbreeding in domesticated stocks of *Catla catla* in Karnataka state and to design a breeding programme for the improvement of the cultured catla stocks.

To this extent the methodologies being used are survey and data collection from state hatcheries for the assessment of the genetic status of the catla stocks presently being used forseed production and culture. The hypothesis of whether the domesticated stocks are inbred due to poor broodstock management will be tested by comparing the level of genetic variation between these and wild stock of catla obtained from rivers, Ganga, Brahmaputra and Mahanadi and by allozyme, mitochondrial DNA (RAPD) and tetranucleotide microsatillite electrophoretic techniques.

Protocols are planned for improving the genetic management of domesticated carp stocks in the state and to disseminate through on-station and on-farm comparative performance testing and by the use of best strains for selective breeding to further enhance their commercial importance.

Progress to Date

A survey of the three main catla producing hatcheries of the state, TBD, BRP and KRP have indicated that the common practices of brood stock replacement seems to be to retain a few fingerlings each year which are normally collected towards the end of the spawning season, consisting of not more than two or three families.

These practices are likely to result in inbreeding and selection for poor growth (Basavaraju *et al.*, 1997). Genetic analysis of the hatchery stocks using four tetranucleotide microsatellite loci has shown these stocks to be different from each other.

Results of growth trials indicated no significant differences in the performance between the stocks of the key hatcheries. However, the interhatchery crosses produced highly significant increases in yield.

The rate of in breeding in smaller hatcheries was found to be unacceptably high which may have brought about a reduction in the culture performance of the hatchery bred seed.

Basavaraju *et al.* (1997) also reported the performance of intra and inter strain crosses between individuals of the same hatcheries viz. TBD, BRP and KRP and between two different hatcheries mentioned above. These studies have shown that crosses between the individuals of the same hatchery did not show any clear differences in growth performance suggesting that no one strain would necessarily be recommended over another.

On the other hand, inter hatchery crosses between the strains at TBD and BRP and the reciprocal cross when stocked under polyculture along with rohu and mrigal at BRP and in two commercial farm ponds and under monoculture in a pond at the Fisheries Research Station (Hassarghatta) have shown an apparent enhancement of culture performance in the crossbred fish over the pure intrahatchery crosses.

There was considerable increase (P<0.001) in yield of the interhatchery crosses compared to intrahatchery crosses ranging from 120–245%, indicating positive heterosis for growth in the interhatchery stocks. Thus, to summarise the findings of Eknath and Doyle (1990) and Basavaraju *et al.* (1997), it is evident that the hatchery practices indicated inbreeding depression and negative selection for growth in domesticated stocks of Indian major carps.

Status of Genetic Research on Indian Major Carps in Other Countries

The most actively involved countries, besides India, on genetic research of Indian major carps are Bangladesh, Vietnam and Thailand. Information is generally lacking regarding genetic work in Pakistan and Burma.

Bangladesh

Research on genome manipulation is being carried out and successful induction of meiotic and mitotic gynogenesis was achieved at the Fisheries Research Institute Freshwater Station, Mymensingh. Clonal lines of *L. rohita* produced by meiotic gynogenesis in the F generation were reported to have been produced. Bangladesh has also initiated selective breeding work on *C. catla* for genetic improvement. For this, catla were collected from different river systems and genetic characterisation is being carried out by applying electrophoresis and morphological methods.

As indicated by the surveys conducted in Indian hatcheries (Eknath and 1990 and Basavaraju *et al.*, 1997), the investigations in Bangladesh have more or less shown the same trend. The existence of 400 carp seed producing hatcheries has been reported from Bangladesh. These hatcheries have been engaged in the production of seed since the 1970s. Some 31 billion spawn are reportedly produced annually. Problems mentioned with regard to hatchery produced seed in India have been also common in Bangladesh. Deterioration of hatchery bred stocks as indicated by retarded growth, morphological deformities and increased incidence of disease has been reported in recent years due to poor brood stock management. The data collected from several fish hatcheries located in Jessore, Comilla and Mymensingh revealed that most of the hatchery operators have no basic idea about brood stock management such as recruitment of new breeders at periodic intervals in the stock, optimum density of the brood stock and their proper feeding (Hussain and Mazid, 1997). These hatchery procedures lead to unconscious negative selection due to the use of undesirable size of brood fish and generation after generation of mating the related spawners from a small population.

Currently, Bangladesh is engaged in characterisation of different land races of catla and rohu by the assessment of meristic and morphological characters, genetic fingerprinting of riverine stocks of catla using molecular techniques and karyotyping to assess ploidy of rohu and mrigal. Developing breeding plans for stock improvement of catla and induction of gynogenesis and production of genetic clones in rohu are part of a genetic improvement programme for carps in Bangladesh.

Thailand

Thailand has an elaborate genetic improvement programme, not only on carps but also for other species such as tilapias, catfishes, freshwater prawns, seabass, oyster, and aquarium fish. Among Indian major carps *L. rohita* has been currently taken up in Thailand for genetic work. Nineteen pairs in 1968 and 250 fish in 1969 were introduced into the country. The species is propagated through hatchery breeding and stocking in reservoirs and swamps where they have been already become established. It is also cultured in ponds. The production of rohu in 1994 was estimated at 1,783 tones ranking third in both weight and value among the carp species cultured in Thailand.

Genetic studies on rohu have been carried out with regard to the growth rate of the species raised in cages and to estimate the heritability as well as the genetic, phenotypic and environmental correlation for body weight, total length, body depth, head length and girth of fish at different ages.

Mrigal was introduced into Thailand during 1980 and 82. A consignment of 100 fish was first introduced from Bangladesh. Later in August 1982 eight pairs of brood fish weighing 700–800 g and 500 fingerlings were brought to Kalasin Inland Fisheries Station. Like rohu the propagation of mrigal is also carried out through hatchery breeding and the seed is stocked in reservoirs and swamps. The species has now become established and is available throughout the country. Thailand wants to set up a National Breeding Programme for hatchery produced carps and tilapias; among the Indian carps, *L. rohita* has been selected. The Central Institute of Freshwater Aquaculture (CIFA) is providing the improved rohu strain under the International Network for Genetics in Aquaculture (INGA) programme.

Vietnam

Vietnam, like India, Thailand and Bangladesh, is one of the member countries of INGA but also the carp genetic improvement programme in Asia. The Indian carp species included in carp culture practices in Vietnam are the rohu and mrigal.

Vietnam has an elaborate selective breeding programme for mrigal. Mrigal production is mainly from north Vietnam, but this species is not considered as important in the south. Of the 200,000 tons production from freshwater, mrigal contribution was reported to be 45,000 tons in north Vietnam. In north Vietnam, sexual maturity in mrigal occurs at two years of age as in India. The breeding process is repeated three times in a season. It has been reported that the first and the second breeding yield better quality seed than the third. This is because young and small brooders normally mature and spawn earlier in the season and, like in other regions, early produced fry fetch higher price. This tendency to prefer to use early mating and small sized brood stock may lead to selection for small size and early maturity.

The base population of mrigal consisted of about 1000 fingerlings, imported from only one source, i.e. the Thangon state fish Farm in Laos in 1984 to the Thu Due Fish Farm of Research Institute for Aquaculture No. 2 (RIA 2) with the support of the Interim Mekong Committee. Another consignment of 10 pairs of breeders were imported from the same source in 1986. At the Fish Farm of the Research Institute for Aquaculture No. 1 (RIA-1) about 4–5 generations have been produced from only 6 males and 6 females which were transported from RIA-2 from the original stock. Because of the repeated production from a limited stock of 6/6 brood fish, the growth rate was observed to decline year by year. Deformities have been consistently observed in about 5–7 percent of the fish. The fish also have been reported to mature at lower age and lower body weight.

Vietnam is planning to take up individual (mass) selective breeding programme for mrigal. However, before the start of the programme, the base population has to be upgraded. Thus, Vietnam has planned to directly import wild populations of mrigal from three different river systems in India, consisting of about 2000 fry from each river. The two main goals of Vietnam for aquaculture of mrigal carp are a high growth rate and a high proportion of females, as males are 60–70 percent of the female body weight at market. This is true in most other carps as well A detailed plan of the selection programme for mrigal in Vietnam has been drawn out by Bentsen *et al.* (1996).

Cryogenic Preservation of Carp Milt and Gene Banking

Protocols have been developed for sperm cryopreservation of major carps more or less simultaneously at the National Bureau of Fish Genetic Resource (NBFGR) and The Central Institute of Freshwater Aquaculture (CIFA) in 1988. The optimum requirements have been developed in respect to extender, activation media, dilution rate, activation period, sperm and egg ratio difference between species. Stocks of *L.rohita* milt cryopreserved in 1988 are being maintained along with other species in a mini gene bank at NBFGR. Viable hatchlings have been produced from five-year old rohu cryopreserved milt (Ponniah, 1997).

The ionic composition of seminal and variation fluid of rohu and mrigal have been studied at the Bureau with the objective of providing basic information that can be used to fashion new extenders and spermatozoa motility inducing media. Gupta and Rath (1993) preserved the milt of *C. catla, L. rohita* and *Cirrhinus mrigala* under cryogenic condition for three years. The quality of milt samples were analysed when preserved and were expressed in terms of spermatocrit value and sperm cell count, which in catla, rohu and mrigal ranged from 65–75, 75–85 and 65–75 respectively while sperm cell counts ranged between 2–2.5 X 10^7, 3.0–3.25 × 10^7 and 2.0–2.5 X 10^7 respectively. The most effective diluent was found to be an electrolyte medium when Dimethyl Sulfoxide (DMSO) acted as cryoprotectant and the ratio of milt and diluent was 1:4. Quick Thawing procedure (38+2°C).

Gupta and Rath (1995) have also studied the efficacy of cryopreserved milt in fertilizing the eggs compared with normal milt.

The survival and growth rates from larvae to fry and to fingerling produced from eggs fertilized with cryopreserved and fresh milt were not found statistically different.

The fertilization efficiency of cryopreserved spermatozoa thawed at different temperatures of catla, rohu and mrigal over a period of four years was assessed at different time intervals after thawing, (Gupta *et al.*, 1995). Milt thawed at 39 + 1C and 30 + 1C had fertilization rates of 56.7% and 12.5% during 900 and 1500 seconds. The motility and fertility were reduced drastically at the next 60 seconds and 300 seconds.

Cryopreservation of Embryos

Cryopreservation of embryos of *Cyprinus carpio* and *L. rohita* have been attempted (John *et al.*, 1993). It has been reported that *C. carpio* at tail bud stage was more tolerant to cryoprotectants than morula stage as revealed by the hatching rates of treated embryos. This may be due to some ontogenic differences in the tolerance to handling, membrane permeability or osmoregulation potential, rather than cryoprotectant susceptibility. Rohu tail-bud embryos treated in methanol and DMSO (Cryoprotectents) in different concentrations and in combination, showed high hatching rates in all the treatments but no pattern was evident. When embryos from all treatments were kept immersed in liquid nitrogen for about 3 hours and thawed, no survival was observed. This was felt due to defective thawing (John *et al.*, 1993).

Cryoprotectant toxicities have been also worked out in two embryonic stages of prawns viz. *Penaeus monodon* and *P. indicus*. Success has been achieved in obtaining viable larvae of *P. indicus* frozen to -40C and of *P. monodon* frozen to -30C. The frozen *P. monodon* larvae developed further without any abnormalities. Also, safe levels of cryoprotectants have been worked out for the embryos of *L. rohita* and common carp. Further research to develop, vitrification and cryopreservation of carp embryos is in progress (Ponniah, 1997).

Gene Banking

The development of effective protocols for cryopreservation of sperm and embryos will go a long way in support of sustainability in aquaculture. It will also facilitate the establishment of a nationa facility for gene banking of fish gametes, embryos and DNA material.

The existence of such gene bank facilities will be also highly helpful for the obvious reasons that it reduces the maintenance problems and costs involved in raising brood fish in *in situ* condition, provides a wider choice of genetic material for selective breeding, and enables the process of dissemination of genetically improved varieties of fish with ease. Conservation of germplasm of endangered and wild populations would become easier.

As mentioned earlier the mini gene bank at NBFGR can be upgraded into a National Fish Germplasm Bank for wild and endangered species. Similar gene banking facilities can be also developed at CIFA for preservation of genetically improved carp and other domesticated culture species.

Breeding Fish

At AC Tpical fish breeding you will finda lot of information and guidelines that can be useful when you attempt to breed your aquarium fish, or when you suddenly realise that your aquarium is filled with eggs or fry. The level of the articles varies from very basic introductory articles to more advanced accounts regarding the breeding of more difficult aquarium species. If you do not find what you are looking for on this page, we suggest that you take a look in the species database as well since a majority of the species profiles contain breeding information.

Some fish will readily spawn in aquariums while others are very difficult to coax into breeding in captivity but this is on of the things thamake fish breeding so interesting. There are also many aquarium species that have never been bred in aquariums and where each specimen has to be caught from the wild.

Fish are ther livebearers or egg layers. Livebearing fish give birth to free swimming fry and the offspring develops inside the female fish. Egg layers will instead release eggs that are fertilized outside the female fish.

Eventually the eggs will hatch and fry will emerge. If you are a beginner aquarist and want to try breedingish, livebearers are usually a better choice than egg layers, since there are several livebearing species that will happily spawn in aquariums even without any extraordinary care or treatment. Molly, Guppy and Platy are three examples of commonlkept livebears.

Livebearers give birth to fry that look like tiny copies of adult fish. It is however common that the fry is protected by a duller and more camouflaging colouration and that they lack extravagant features that are found in mature fish.

During the breeding period, the male will use his gonopodium to fertilize the eggs inside the female. The gonopodium is actually the anal fin of the male fish that has developed into a reproductive organ. After fertilization it will typically take four weeks before the fry is born. The fry is not born until it is developed enough to survive without any help from the parents.

Different egg laying species have developed different breeding strategies. Some are devoted parents that build advanced nests for their offspring or even carry them around inside the mouth, while others will simply release the eggs and let the current sweep them away as soon as they have been fertilized. Marine fish species will often release eggs that can float and stay hidden among plankton.

Egg laying species are usually divided into four basic groups: egg scatters, substrate spawners, bble nest builde and mouth brooders. Fish species that live in rivers and streams are often egg scatters and they will typically let the current transport the eggs away and distribute them over a large area.

The egg can also fall down to the bottom where the fry develop. Since egg scatters hardly ever encounter their offspring in the wild, they are likely to eat them in the aquarium since the enclosed space is an unnatural living environment. Substrate spawners will instead attach the eggs to some type of substrate to make it impossible for currents to carry them away.

Some species will only use one specific type of substrate, and sometimes also vigorously clean it beforehand, while others are less selective. It is common for the male and/or the female to create a form of nest in the substrate. Bubblenest breeders will instead create a nest that is floating far away from the substrate. In most bubblenest building species it is the male that is responsible for building the nest and he will blow out tiny bubbles that he carefully glues together with siva. Mouth brooders carry eggs and/or fry inside the mouth of a parent. Some species are maternal mouthbrooders while others are paternal mouthbrooders. Mouthbrooding will often be combined with substrate spawning.

Breeding Angelfish

Sexing Angelfish and Obtaining a Suitable Pair

It is usually very hard to distinguish a female Scalare form a male one, since the look very similar. Most Angelfish keepers are actually unsure of the sex of their fish until they notice which one is laying the eggs and which one that is fertilizing them.

If you watch your Angelfish closely right before the spawning commence, you will however notice a small difference in the size of the papilla between the sexes. The papilla is a pink coloured organ formed between the anal fin and the ventral fin.

Both sexes will display a papilla right before spawning, but the Angelfish with the largest papilla is usually the female one. Her papilla will also be blunter than the males, and when the female Angelfish becomes gravid she will display a large bulge in the region around her papilla. You can also notice a difference in behaviour; a male Angelfish is more territorial than a female. If one of your female Angelfish becomes gravid, all the male Angelfish in the aquarium will form a papilla. Those fishes that do not display a papilla in this situation are either females or unhealthy males.

Angelfish forms pair before breeding. They do however not pair up for life. The easiest way of obtaining a pair is to buy an already established pair from a breeder or a fish store.

This is however quite expensive and forming your own pair is much cheaper. A recommended way of forming a pair is to buy 4-8 Scalare and place them in the same aquarium, where they can get to know each other and form their own pairs. Choose fish with good external features, that looks strong and that display a healthy appetite.

Study the fins and gills extra carefully for signs of illness. It is advisable to buy from an established breeder if you want to produce high quality Angelfish.

Getting Angelfish into Spawning Condition

Your Scalare will usually need to become at least 5 cm / 2 inches before they are mature enough to spawn for the first time. If kept on a suitable diet, this will happen when the Angelfish is between 8 and 12 months old. To induce spawning, you should keep your Angelfish in an aquarium that is large enough and not overcrowded.

The aquarium should ideally be at least 40 cm / 16 inches high, since the Angelfish is a tall fish. An aquarium that is to shallow will stunt their growth.

Angelfish can survive quite poor water qualities and unsuitable water temperatures, but it can make them stop eating and will inhibit breeding. Frequent water changes should therefore be performed and the water temperature kept between 80 and 85 degrees F (27 and 29 degrees C). Suitable food is also necessary to get your Angelfish into spawning condition. The diet should be varied. Prepared foods especially made for Angelfish is a good base, but should be supplemented with meaty foods. Mosquito larvae, daphnia, beef heart, brine shrimp and larger brine shrimp cut up into suitable size are food types known to induce spawning in Angelfish.

Even though your Angelfish must be given plenty of food, you must also avoid overfeeding them. Check the aquarium 5 minutes after the feeding and remove any food that has not yet been consumed. If you frequently find left over food in the aquarium after 5 minutes, you are feeding your Angelfish to much.

Spawning Signs in Angelfish

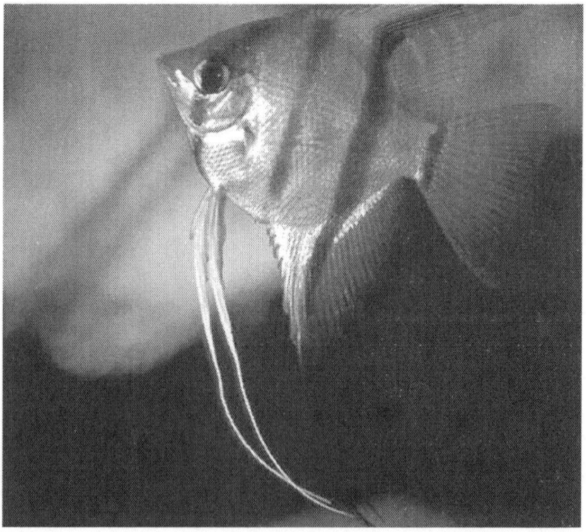

Figure: *Choose good parents when breeding angelfish.*

If an Angelfish suddenly displays a bulging belly and a more aggressive behaviour, it is probably a gravid female Angelfish. Two Angelfish grooming each other is also a typical breeding behaviour. Angelfish of both sexes will also flash fins, face off, lock their mouths and twist around. When a couple has been formed, they will choose a spawning site and start cleaning it together.

Angelfish Egg Laying and Fertilization

The Angelfish eggs will be deposited on the spawning slate and placed in very neat and evenly spaced lines. If she doesn't like the spawning slate, she will deposit the eggs somewhere else in the aquarium. The male Angelfish will fertilize the eggs by following close behind the female and touching all the eggs with his papilla. If you keep only female Angelfish, one of the females might actually resume a male behaviour and follow the egg-laying female and touch the eggs. Those eggs will naturally remain unfertilized and become white within a few days.

Angelfish Egg and Fry Care

Several methods are used by Angelfish breeders to protect eggs and fry from hungry adult fish. Some remove all fish except the breeding pair from the aquarium, while other breeders choose to remove the breeding pair instead and place them in a special breeding aquarium. Some remove less dominant males until only the most dominant male is left with the females in the aquarium. This method also involves removing all non-gravid females from the aquarium.

All adult Angelfish pose a treat to the eggs, and many breeders will therefore place the eggs in their own tank as soon as the eggs have been fertilized. The easiest way is to put a special spawning slate in the aquarium before the spawning takes place. The parent fish will deposit the egg on the spawning slate, and the spawning slate can be easily moved afterwards. A clay pot, poly pipe or a piece of smooth surfaced tile will all work as spawning slates. The eggs will survive a brief moment out of the water when you lift them from one aquarium to the other, but try to make the move as fast as possible.

If you do not wish to keep all your fry, you can let the eggs stay in the aquarium with the adult fish. If you provide your fry with suitable hiding places, such as heavily planted areas, at least a few of them will reach adulthood. Be aware of strong filtration, since small fry can be sucked into filter systems.

If your Angelfish pair has never spawned before, it is usually a good idea to let them spawn once or twice in their normal aquarium "on their own" without your meddling. If you move them to a breeding aquarium during their first spawning they might be confused and stop the spawning process. It is normal for the parents to eat the eggs from their first spawning, and if this happens the female will usually deposit eggs again after just a few weeks.

Angelfish Development

Figure : *Angelfish breeding have result in many types of angelfish*

Newly deposited Scalare eggs are transparent or semi-transparent, with a pale yellow colouration. If the eggs are not fertilized, they will start turning white after approximately 24 hours. If they have not been eaten by other inhabitants of the aquarium, unfertilized Scalare eggs will be attacked by fungus and start looking very fuzzy after roughly 48 hours. 48 hours after being deposited, fertilized eggs on the other hand will begin showing wiggling fry tails.

On day three, tadpole shapes with large yolk sacks will be visible. The tadpoles will still be attached to the breeding slate. If the parent Angelfish are still in the same aquarium as the egg, the parents will now start to move the small fry around. On the fourth day you will be able to see tiny eyes on the fry and on the fifth day the yolks will start shrinking. Very soon the first fry will be free swimming. During day seven the fry will usually start getting hungry since their energy from the yolk have been consumed.

Suitable Tank for Angelfish Fry

If you decide to raise the Angelfish eggs and fry in a special aquarium, a 5-10 gallon (19-38 litres) tank will be suitable, but even a tank no larger than 2 gallons will usually work. You do not need to cover the egg and fry aquarium. If you allow the parent Scalare to accompany their offspring in the aquarium, the parents will keep the eggs clean and protect them from mould.

If the eggs are kept without the parents, you must use fungicide to prevent fungus attacks. Make sure that the water temperature in the raising tank is the same as in the parent's old aquarium. An air stone in the raising aquarium will make it easier to maintain high oxygen levels. Keep the water quality good by frequently removing any dead fry and left over food. Regular water changes are also a must. Typically, a lot of the Scalare fry will die. You might have to remove dead fry from the aquarium three times a day or more. During the first three weeks, water changes can be necessary as often as once a day.

Feeding Angelfish Fry

You can start feeding your Angelfish fry around 7 days after the eggs were deposited. Angelfish fry can survive 12 hours without food, but should ideally be fed four or five times a day. Newly hatched Brine Shrimp (Artemia nauplii) are a good first food for Angelfish fry. If the fry is too small to consume the Brine Shrimp, you can buy the fry special fry food from a pet shop or fish store. The survival rate for Angelfish fry is usually very much lover when the fry is fed dead food instead of live food (such as live Brine Shrimp). When the fry are a few weeks old it is time to make them grow accustomed to flake foods. Turn the flake food into a powder or press it through a fine screen before feeding it to the fry.

Angelfish Guarding and Caring Behaviours

As mentioned above, it is quite common for Angelfish to eat their eggs or fry, especially during the first few spawnings. This is why most Angelfish breeders place the eggs in a special aquarium without any adult fish. It is however very fascinating to watch an Angelfish pair taking care of their own offspring. You can for instance remove some of the eggs, and let some stay with the parents. If you leave the eggs to hatch with the parents, the parents will spawn again after 2 or 3 weeks and the already hatched fry will use the eggs as food. When the fry is 5 or 6 weeks old you will see them eat from the skin and fins of their parents. This behaviour can make the parents very weak and even cause hem to die.

Some Scalare individuals are more aggressive than others, and will not be able to raise thei own offspring. They will always eat their offspring or at least chaise them away. Such individuals can usually be singled out as early as during the courtship period, since they are extremely aggressive and sometimes even injure a fish they wish to breed with. For instance, it is not that rare for Angelfish to suffer from severely injured mouths after the courtship period.

Breeding Bristlenosed Catfish

Bristlenosed catfish is not very hard to breed in aquariums. Keep one male Bristlenosed catfish with several females, and make sure they are all in top condition. Provide them with good water quality and a varied and nutritious diet. You will most likely never notice the actual spawning behaviour, just the result of it in the form of tiny Bristlenosed catfish babies that suddenly appear in the aquarium.

From what we know about Bristlenosed catfish breeding habits, the spawning seems to be initiated by the male catfish. He will pick a suitable spawning site, usually a cave or a crevice, and subsequently start chasing the females as a part of the courting behaviour. The females will repeatedly inspect the spawning site and this part of the spawning can go on for several days. Eventually the female catfish will deposit her eggs at in a clump at the spawning site where they are ertilized by the male. Newly deposited eggs from a Bristlenosed catfish display a bright orange colour. The female will leave the spawning site immediately and do not stay around to guard or care for the eggs.

he male Bristlenosed catfish stays at the spawning site to protect the eggs from predators. The will place his tail near the eggs and let his head protrude out of the cave or crevice. He is probably placing his tail close to the eggs in order to care for them by fanning fresh water over them, but this has still not be determined. The male Bristlenosed catfish is a very capable parent and you should never remove him from the eggs or try to raise catfish eggs in a breeding tank without the male present.

Fry will emerge from the eggs after 4-5 days. The fry will still eat from their yolk-sac. They will start to spread over the aquarium, and in a community aquarium a lot of them will soon be consumed by the adult fish. If you want to ensure a high fry survival rate it is recommended to breed Bristlenosed cafish a separated tank. You can even remove the females from the aquarium, and leave the fry with the male. When the fry has been free swimming for a week you can remove the male as well. High oxygen levels are very important for Bristlenosed fry and good aeration in the aquarium will typically lead to a higher survival rate.

When the yolk-sac is consumed you can start feeding the Bristlenosed fry vegetable food, such as very finely cut lettuce and cucumber. Dip the vegetables in boiling water before you place them in the aquarium. A good alge growth in the aquarium is beneficial for the fry since this is their natural food source. If kept on a suitable diet the fry will be around 1 ¼ inches long after two months.

Breeding Discus Fish

Discus is not a very easy fish to keep, and breeding them in aquariums is even harder. Only those with previous experience from keeping easier species should try caring for Discus, since many Discuses have died in the hands of inexperienced aquarists. Discus fish are wonderful to keep – they are beautiful as well as charming – but they do require quite a lot of pampering. The key to successfully breeding Discus lies in always providing them with supreme water conditions, keeping them in an aquarium that is large enough for them, giving them high quality food etcetera. A well kept Discus pair will breed regularly in the aquarium, and there is really no special "fix" that will induce breeding in a couple that do not feel well. Most of the guidelines regarding Discus breeding will therefore be the same as for keeping Discus.

First and foremost – a Discus couple will not breed in an aquarium that is too small. A depth of at least 15 inches is required for these tall fishes. A 36 inches x 18 inches x 18 inches aquarium is therefore usually the smallest suitable standard aquarium for a pair. If you have a 48 inches x 18 inches x 18 inches aquarium you can even keep a group of 4-6 smaller Discus until they have formed their own pairs.

The water temperature should be kept at 82 degrees Fahrenheit or higher, and the water must be very soft. Keep the pH as close to 6.5 as possible. The soft water means that you will have to monitor the pH value vigilantly. Soft water is a very bad buffer and the pH value can therefore drop very quickly and your Discus will not do well in acidic water. Keep the levels of soluble waste as low as possible and never allow the nitrate level to rise above 20 ppm. Since your

Discus will require a diet rich in protein, they will consequently produce a lot of waste products that will push up the nitrate levels in the aquarium. The best way to combat the nitrate is by performing frequent water changes. Changing 30-50 percent of the water each week is a good rule of thumb.

As mentioned above, your Discus will require plenty of protein in their diet. Professional Discus breeders usually give their fish beef heart. The beef heart diet should ideally be supplemented with other food types, such as spinach and bloodworms. Some high quality flake food will provide your Discus with extra nutrients, and additional vitamins can also be a good idea.

During the initial stage of the breeding process your Discus couple will start to defend a small region of the aquarium. Both the male and the female Discus will become highly territorial and guard this region together. They will continue by picking a spawning site and carefully clean it. Some Discus pairs clean a number of sites before they finally decide to place the eggs on one of them. The female Discus will swim over the spawning site and deposit her eggs in rows, and they male will swim behind her and promptly fertilize them. I wish to once again stress the importance of keeping your Discus in soft water. Hard water will make the protecting membranes that surround the eggs so hard that it is impossible for the sperm to get through. Hard water can therefore cause the eggs to stay unfertilized and no fry will develop. Hard water can also make it impossible for a fry to hatch.

Discus pairs are attentive parents nd will guard and care for the eggs together. They will fan fresh water over the eggs to ensure high levels of oxygen and they will also remove any unfertilized eggs to prevent bacteria growth and mould. Discus eggs will hatch within 48 hours if the water temperature is right. The fry will stay at the spawning site with the parents for an additional 12 hours. After that, the fry will be moved around the aquarium by their parents, and the parents will use different places to hide the fry in. If you keep up the water temperature your will have free swimming fry in the aquarium after approximately one week. Free swimming fry will need no extra food, because they will eat from the scales of their parents. The adult Discus will secrete baby-food from between their scales. This food will provide the fry with everything they need for about one week. After that, they will be large enough to eat more normal food and you can start feeding them newly hatched brine shrimp. Powdered flake food is another option. Just like their parents, Discus fry will appreciate some beef heart, but since the fry are so small you will need to liquidise the beef heart for them. Do not separate the fry from the parents until you know for sure that they are no longer feeding from the scales. If you provide your Discus fry with ideal conditions and suitable food, they will grow very rapidly and be around 2 inches in diametre at an age of 12 weeks. When they have reached this size, they are large enough to be sold to pet shops or other aquarists.

The parents are extremely important when it comes to keeping eggs and fry alive and it will be practically impossible for you to raise Discus without the parents. Do not be discouraged if you Discus pair fail to raise their first batch successfully. The mere fact that they are spawning in the aquarium is a good sign, a one failed spawning will usually be followed by a more successful one eventually. It can take a Discus couple numerous spawnings before they finally get it right.

Breeding Goldfish

Goldfish cover a wide range of species, but they are all members of the generic family Carassius auratus, and have been bred by the Chinese, Koreans and Japanese. One of the difficulties for the amateur to breed this fish is that they can be difficult to sex. The difference in their gender is only apparent when they are in a spawning condition. The greatest change occurs in the males, they develop white pimples known as "breeding tubercles", on their fins and sometimes their head. Check along the gill operculum, and along the front rays of their pectoral fin.

Goldfish are capable of breeding when they are a year old, but they are in their prime during their third year. Under normal circumstances, it is impossible to tell their sex until they have grown to 8cm. Therefore it is impossible to buy a small fish and be certain of its gender. Many breeders advocate that the sexes should be kept apart before the breeding commences, and this is impossible before they are sexed. Therefore breeding of goldfish can be a long term project, as they first have to be sexed. This is best done in the Summer, and the fish can be over wintered for breeding in the Spring.

The process of breeding is induced by a change in the temperature of the water. Goldfish will breed at a temperature of 10-26°C, but the optimum temperature to ensure success is 20°C. To induce the fish to spawn you can either wait for more clement weather, or you can artificially and slowly raise the temperature of the water. To ensure a greater chance of fertility more males are kept then females, usually the ratio is two to one. Bear in mind here that each fish needs fifteen gallons of water, and they do not like to be overcrowded.

It is easy to spot when they are ready to breed, as the male has a ritual of a "spawning chase", whereby he follows the female around and pushes against her abdomen, for several hours. This should occur if all the conditions are correct with a few days of the fish being brought together. The female, when she is ready will lay her eggs in batches, the first batch being the most fertile. She can lay from five hundred to a thousand eggs, so they will not be released all at once. The male will immediately fertilize these eggs by spraying milt on to them.

Once the egg laying is over the eggs should be removed from the breeding tank, and rinsed in aged water, which isat the same temperature as the water they have been released from. They should then be placed in a separate twenty gallon tank. However the depth of the water should not exceed six inches, as the pressure of the water inhibits their ability to rise to the surface.

The eggs will hatch at the temperature of 21°C, so if the water is cooler raise it gently to that temperature. At this stage the clearer looking of the eggs are the most fertile, and therefore have the greater chance of hatching. The darker eggs are infertile, and have a high chance of developing a fungal infection; they should be removed from the tank. A few drops of methylene blue can reduce the chance of fungal activity.

Figure: *Goldfish breeding have resulted in many types of goldfish over the years*

Once the fry have been hatched they have a full yolk sac, and they should not beefed during the first forty eight hours. During this time it is normal that they will sink to the bottom of the tank, and they should be left undisturbed. During this period they are absorbing the yolk sac and developing an air bladder. Once the air bladder has developed, they will rise from the bottom naturally.

After forty eight hours they are hungry and ready to be fed. They may be fed from a commercial product or from a paste of oatmeal, and also hard boiled eggs. They need only sufficient food, which they can eat during a twenty minute period, after this time any uneaten food should be carefully siphoned off.

Breeding Mbuna Cichlids

Mbuna cichlids are native to Lake Malawi in Africa. The word Mbuna means rock dweller in one of the local languages spoken around the lake, and it is truly a suitable name for these rock living cichlids. Mbuna cichlids are popular in aquariums due to their beautiful look – most species display very striking colours – and their fascinating behaviour.

They are very aggressive and should not be kept with more timid fish. Mbuna cichlids will claim territories and defend them violently. They should ideally be kept in a well stocked aquarium, since this will cause the aggression to be spread out over many individuals. One inch of fish for every 8 square inches of water surface is a good rule of thumb when it comes to Mbuna cichlids.

It is not hard to get Mbuna cichlids into spawning condition. Just provide them with the same care and conditions as you would provide non-breeding Mbuna cichlids with. This article will provide general guidelines that are true for most Mbuna cichlid species, but it is always advisable to research your particular species as well before you attempt to breed them.

Also remember that Mbuna cichlids from different species can and will cross breed. If you keep several species together and they produce offspring you should not sell the fry since the mixed DNA will ruin the bloodlines of the true species when people use the crossings for breeding purposes.

Keeping one male with at least 2-3 females is a good ratio. As mentioned above, the Mbuna cichlids should be kept in a well stocked aquarium. This means that you might have to add extra aeration and filtration, and perform water changes quite often. Your Mbuna cichlids will not breed if you allow the water quality to trop. Lake Malawi is a large lake with a stable environment, and Mbuna cichlids to not appreciate rapid changes. Keep the pH in the 7.5-8.5 range and the KH as well as the GH at 12.

The levels of ammonia and nitrite should be zero, and the nitrate should never rise above 15 ppm. The normal temperature in Lake Malawi is 75-80 degrees Fahrenheit.

If you want to breed you Mbuna cichlids, it is important not to over feed them. They will eat virtually everything you give them and can easily become obese which is bad for their health. Very large Mbuna cichlids also means that you will be able to keep fewer individuals in your aquarium, and the amount of aggressive behaviour will therefore increase. Violence will make the fish stressed and they are less likely to breed under such conditions.

In the same time, it is important not to overcrowd the aquarium, since this will hamper the breeding as well. The natural diet for Mbuna cichlids are algae and insects. If you feed you Mbuna cichlids prepared food, you should always include some vegetable/algae matter such as Spirulina.

Since Mbuna cichlids are rock dwellers, it comes as no surprise that they like to spawn among rocks, caves and crevices. Mbuna cichlids should never be kept in an aquarium without plenty of hiding spaces created by rocky decorations, and if you want them to breed this is even more important.

During the initial stage of the breeding period the male will claim a small territory which he thinks is a suitable spawning site. He will then start to show him self in front of the females. He will 'dance' for them by vibrating his whole body and erecting his fins. His colours will also become much more vivid than normally. If a female is in spawning condition, she will eventually accompany the male to the spawning site.

The couple will swim close to each other in a circling fashion, and the female will start depositing eggs at the spawning site. When a few eggs have been deposited, she will quickly pick them up with her mouth to guard them. Mbuna cichlids are femalmouth brooders and will keep their offspring hidden inside the mouth of the mother.

The male is equipped with some egg-shaped spots on his anal fin, and when the female Mbuna tries to pick them up as well she will instead receive a mouthful of sperm that fertilizes the eggs inside her mouth. She will then continue to deposit eggs at the spawning site, and the procedure is repeated over and over again until all the eggs are fertilized and resting inside her mouth.

When all eggs has been fertilized, the male Mbuna will chase away the female from his territory. It is unadvisable to remove the female from the aquarium at this stage, since the stress from being netted and moved can cause her to spit out the eggs. If you have filled the aquarium with a lot of small hiding places for the fry to seek refuge in, you can let her stay in the aquarium and let the fry grow up together with the adult fish. Some fry will be eaten, but at lest a few will survive. Their instinct will make them seek shelter among rocks and stones just like they would do in the wild. If you want to ensure a higher survival rate, you can move the female to her own aquarium approximately 7-10 days after spawning. She is less likely to spit her eggs when she has kept them in her mouth for such a long time.

The fry is usually spit out after 4 weeks of brooding, but you should not move the female later than 10 days after the spawning since a late move can hurt the fry.

When the Mbuna fry are released they are already comparatively large and you can feed them Cyclops, brine shrimp or powdered flake food. One batch of Mbuna fry will typically include 25-35 individuals. Let the female stay with her fry for a while and give her plenty of nutritious food. She needs some time to regain her strength before you move her.

Breeding Oscar Fish

Oscar fish can be bred in aquariums as long as you provide them with favourable conditions. To begin with, the aquarium must be large enough. Oscars are large cichlids and their size can exceed one foot. To get a breeding pair, you must house 6 young Oscars together and let them form their own pairs. Make sure you choose healthy Oscars from different broods. It can be hard to sex young Oscars, but you will hopefully get at least one pair from 6 individuals. If you don't want to wait and let Oscars grow up together, you can instead buy a breeding pair but this is usually much more expensive. The stress and disturbance experienced by the fish when you move them from their old aquarium can also cause them to brake up and stop being a breeding pair.

Once a pair has been formed you should remove the other fish from the aquarium, since breeding Oscars are highly territorial and can hurt intruding fish. If the spawning does not commence immediately there is not easy fix that can kick start it. What you can do is provide the pair with suitable conditions and wait for the spawning to start.

The courtship behaviour can be very violent – sometimes too violent and resulting in the death of one of the Oscars before any offspring has been produced. Oscars will do well in soft acidic water as well as in harder alkaline water, as long as you avoid any extremes. Keeping the conditions similar to their native home in Central America is always a good idea. Frequent water changes are important to ensure suitable water chemistry and low levels of soluble waste. Do not change more than 25 percent of the water at once. Changing 20-25 percent of the water once or twice a week is a good rule of thumb. Efficient filtration is also required since Oscars are renowned for being messy eaters and producing plenty of waste. If you have a 50 gallon aquarium you should ideally use filtration that is recommended for a 100 gallon aquarium.

Feed the pair bait fish or frozen fish. Prawns, crickets, earth worms and other meaty foods are also suitable. Some Oscar keepers feed their fish raw meat from cows and pigs, but this is highly unadvisable since such meat include too much animal fat and can harm the health of your Oscars.

Remember that wild Oscars feed mainly on small fish and eating cows, pigs, sheep etc are therefore very unnatural for them. Pellets should also be avoided since a majority of the pellet will slip through the gills of the Oscar and never end up inside the stomach. Your fish will still be hungry and the water will be polluted with uneaten pellet material.

An Oscar pair will typically dig a pit in the substrate or choose a hard surface, such as a flat stone, as their spawning site. Both the male and the female Oscar will guard and care for eggs and fry.

The will be extremely protective and aggressive during this period and should be kept alone. Just like with many other cichlid species, it is common for an Oscar pair to fail their first attempts to raise offspring.

Newly hatched brine shrimp is a suitable first food for Oscar fry. You can feed the powdered flake food, but they will grow much faster on brine shrimp. When kept on a brine shrimp diet the fry will be 1.5-2.0 inches long after three months. Keep in mind that one Oscar pair is capable of producing tremendous amounts of offspring. You might have to cull the batch if you can't provide all the young Oscars with suitable housing.

Breeding Tiger Barbs

Tiger Barbs are a good choice for beginner aquarists that wish to breed fish, since Tiger Barbs frequently spawn in aquariums without much pampering. They are hardy and will accept a wide range of temperatures. If you choose a high quality flake food they will do well on a prepared food diet, but it is always recommended to sometimes supplement the flakes with more meaty foods such as Daphnia, bloodworms or similar.

The Tiger Barb is an active fish that can reach a size of 1.5", so make sure you have enough space to keep the produced fish in before attempting to breed them. Since Tiger Barbs do not raise their offspring, one Tiger Barb batch is usually very large to make sure that at least a few of the fry survive. This means that you might have to cull the batch if you do not have a very large aquarium for the fry to grow in. If you refrain from culling and instead let them grow up in a crowded aquarium, the whole batch will suffer.

If you want to breed Tiger Barbs, you must start with separating the males from the females. It is true that healthy Tiger Barbs sometimes spawn when kept together in community aquariums, but this it is not something that happens very often and if you want to induce spawning you should separate the sexes from each other. The smallest Tiger Barbs with the plumpest bodies are the females.

When you have separated the sexes you should start feeding them a spawning-inducing diet. Tiger Barbs are happy eaters and will usually consume anything you give them, but breeding Tiger Barbs require a diet rich inprotein to do well. Freeze dried Tubifex is known to induce spawning in Tiger Barbs, but other protein ric food types can also work well. Live food is always a good idea.

The temperature in the aquariums where you keep your Tiger Barbs should stay around 80 degrees Fahrenheit. After 3-4 days of separation you should let the Tiger Barbs see each other again. The aquarium where you want them to breed must have slightly soft water, and the pH should be a bit acidic.

Just like when they were separated, the water temperature should be kept close to 80 degrees Fahrenheit. A mature sponge filter will help you keep the water conditions optimal. You must also provide your Tiger Barbs with suitable spawning sites, such as two bunches of Cabomba plants or a pair of spawning mops.

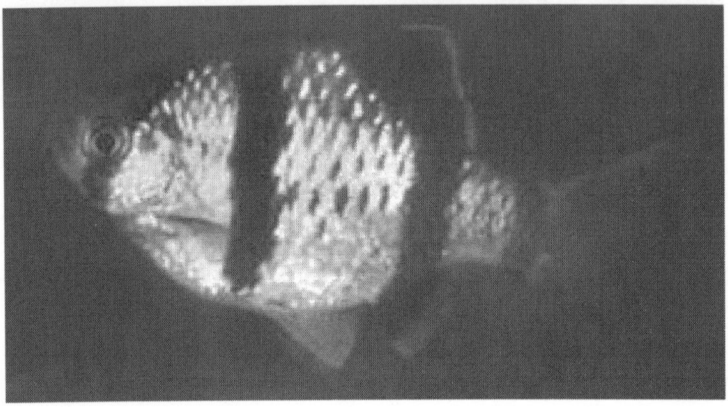

Figure: *A Tiger Barb in breeding condition.*

Spawning will usually begin very soon when the Tiger Barbs have been brought together again. During the courting period, the male will chase the females. When a female Tiger Barb starts releasing her eggs at the spawning site the male will follow her and leave a trail of sperm in the water.

When the spawning is over you should remove the adult fish from the rearing aquarium, since Tiger Barbs are known to eat their own offspring. If you have covered the aquarium floor with marbles the eggs will be safer since the eggs can fall down between the marbles where the adult fish can't reach them. If you keep the water temperature in the aquarium at 80 degrees Fahrenheit you can expect free swimming fry after approximately five days. The fry will usually cling to the sides of the aquarium.

Newly hatched brine shrimp is a suitable first food for Tiger Barb fry. Powdered flake food will also work, but the growth rate will be slower and powdered flake food is not recommended for larger Tiger

Barb fry. If the newly hatched Tiger Barbs are very small, you can feed them liquid fry food until they are big enough to eat solid food. Tiger Barbs are commonly sold to pet stores when they are around three months old.

Breeding/Rearing Aquarium Tanks

Some fish species will successfully breed in a community aquarium or species tank, but others will require a separate breeding aquarium. In some cases, the fish will spawn their normal aquarium but the fry must be caught and placed in a separate rearing aquarium to prevent the fry from being eaten by the adult fish. Fry left with adult fish will usually be consumed, but you can increase the survival rate by providing the fry with plenty of hiding spaces among marbles, densely planted bushy plants etcetera.

If you decide to setup a breeding/rearing tank, a 2-6 gallon container will be large enough for most aquarium species. Very large fish might require 10-22 gallons. The easiest way to fill your breeding tank with suitable water is to fill half the tank using water from your large aquarium where the adult fish live. Use clean water to fill up the rest of the breeding tank.

This water should be treated in the usual way for water changes. Another method is to place a foam filter in your large aquarium and let it run for a few weeks before you set up the breeding tank. This way, the foam filter will pick up a lot of necessary bacteria and the biological process in your breeding tank will get a kick-start when you insert the foam filter. A third method is to buy commercially available products such as Bio-Starter or Safe Water from your fish store and use them in the breeding tank. Some breeders combine the methods described above.

It is not recommended to place a lot of decorations in the breeding tank, unless you know for a fact that your particular species require some special type of decoration. A thin layer of gravel is easier to keep clean than a thick one, and cleanliness is very important in a rearing aquarium. A foam filter is the best type of filtration in a breeding tank since the filtration is gentle but efficient. The air pump must be of correct size for your breeding tank, or an adjustable model.

Even with a good filter, regular water changes are absolutely imperative in a rearing tank. High levels of soluble waste can harm and even kill eggs and fry. If there are no fry or adult fish in the aquarium, water changes can be performed much less frequent since eggs hardly produce any waste products. When the fry is very small, siphoning out waste products and left over food using a flexible tube is usually the safest cleaning method. Fish fry are renowned for their curiosity and accidents are common. Let the other end of the flexible tube rest in a clean bucket (do not clean with soap, detergent or similar!) when you siphon. This way, you can rescue the fry that becomes siphoned up and put them back in the rearing aquarium.

Plastic breeding traps are popular to separate fry from adult fish in community aquariums and species aquariums, but are not suitable homes for the fry for any length of time. The circulation of water is usually far from optimal and the fry should therefore be moved to a rearing aquarium as soon as possible. Another problem with plastic breeding traps is that many models float directly under the aquarium lights, which makes the water inside unsuitable warm for the fry. A breeding net that can be attached to the side of the aquarium is a better solution if you can't setup a separate rearing tank for the fry.

3

Fertility of Interspecific Hybrids

Almost all the interspecific hybrids matured within the third year of age.

Table: List of interspecific hybrids between Indian major carps

Female, Male	*Hybrid*	*Comments*
1 *L.rohita* × *L.calbasu*	(rohu-kalbasu)	Hybrid grows faster than Kalbasu and fertile.
2. *L.calbasu* × *L. rohita*	(kalbasu-rohu)	Hybrid grows faster than Kalbasu and fertile.
3. *L.rohita* × *L. bata*	(rohu-bata)	Hybrids exhibited poor hatching and slow growth rate.
4. *L.calbasu* × *L. bata*	(kalbasu-bata)	Hybrids exhibited poor hatching and slow growth rates.
5. *L. gonius* × *L.calbasu*	(gonius-kalbasu)	Hybrids exhibited poor hatching and slow growth rates.
6. *L. fimbriatus* × *L.rohita*	(fimbriatus-rohu)	Hybrids grow faster than fimbriatus and have better feeding efficiency.

Production of F2 Generation of Interspecific Hybrids

The matured kalbasu-rohu hybrid has been bred through hypophysation and the F2 generation was produced successfully in 1960. As in the case of the F1 progeny, the percentage of fertilization in the F2 progeny was also recorded to be high. The hybrids possessed varying characters which were intermediate between the original parents viz. *Labeo calbasu* and *Labeo rohita.*

Intergeneric Hybrids

Of the 13 intergeneric hybrids produced only four hybrids, *L. rohita-C. catla, Cirrhinus mrigala-C.catla, L.rohita-Cirrhinus mrigala, L. fimbriatus-C. catla* and the reciprocal hybrids have been found to possess useful traits in terms of growth (Ibrahim, 1977; Basavaraju and Varghese, 1980 b). The following are the characteristics of these hybrids.

Characteristics of Some Useful Intergeneric Hybrids

L. Rohita × C. Catla (Rohu-catla) Gybrid

Phenotypically the hybrid exhibits intermediate traits to that of the parents, but tending more towards the maternal parent (L.rohita) with regard to body proportions and fin ray counts. Colouration of the body of the hybrid was near to catla while the shape of the mouth was like rohu. However, its position with infringed lips was terminal and devoid of barbels. The hybrid possessing a smaller head than that of catla and wider body than rohu shows improved traits over the parents. The flesh content of the rohu-catla hybrid, 54–58%, has been also reported to be the highest when compared to rohu with 48–52% and catla with 45%.

Growth

With regard to growth traits, the hybrid did not differ in growth from its parents in the initial stage. However, in the grow-out stage the growth rate of the hybrid has been reported to vary. Chaudhuri (1973) observed the growth rate of this hybrid was like that of catla, while Chaudhuri *et al.*, 1975 reported, it to be 718–915g in one year duration in carp polyculture experiments, whereas Bhowmick *et al.*, 1981 observed it to be 900–1035g within the same duration. Other workers, (Varghese and Shantharam 1979, Keshavanath *et al.* 1980, Konda Reddy and Varghese 1980, Varghese *et al.* 1984, and Somalingam *et al.* 1990) have also reported that the rohu-catla hybrid grows better than rohu. Konda Reddy and Varghese (1980b) conducted experiments on the growth performance of the rohu-catla and catla-rohu hybrids along with the parent species. They observed that the rohu-catla hybrid (female rohu and male catla) showed slightly faster growth than rohu but very much slower than catla. Keshavanath *et al.* (1980), also evaluated the growth performance of the hybrids produced between *L. rohita* (female) × *C. catla* (male) and the reciprocal hybrids along with their parental species under communal rearing and also independently. These studies have shown that the growth of both the hybrids was inferior to catla and almost equal to rohu. The growth difference was insignificant between the hybrids. However, when reared separately, the growth performance of the hybrids was better than rohu but inferior to catla.

Varghese and Shantharam (1979) made a comparative study on the growth of three intergeneric hybrid of Indian major carps viz., hybrids produced by crossing female rohu × male catla (rohu-catla), catla female × male mrigal (catla-mrigal) and rohu female × mrigal male (rohu-mrigal). After 90 days of rearing, the rohu-catla hybrid showed the fastest growth among the three hybrid types, followed by the catla-mrigal hybrid. Whereas, the rohu-mrigal hybrid attained the lowest growth of all the three. In terms of percentage increase in the growth the rohu-catla and the catla-mrigal have respectively shown 257.6% and 110.9% more over the rohu-mrigal hybrid.

According to the observations of Chaudhuri (1973) among the Indian carp hybrid rohu-catla was the best as it combined the quick growth of catla and small head of rohu. He reported that the growth of this hybrid was faster than rohu and nearly as fast as catla. However, the observations of Konda Reddy and Varghese (1980 b) have shown that this hybrid grew slightly faster than rohu, but its growth when compared to catla was very poor. Somalingam *et al.* (1990) studied the growth of rohu-catla hybrid stocked in three different water bodies, i.e., ponds and small and larger reservoirs. The growth rate of the hybrid in ponds when compared to the parent species was 10–20% less than the male parent catla and 30–125% more than the female parent rohu. However, in small reservoir the growth of the hybrid during the second year of rearing equalled to catla though it was less by 20% and 32% during the first and third years of rearing. At the same time, the hybrid has shown 53%, 128% and 32% higher growth over rohu during the first, second and third year respectively. The authors suggested that the rohu-catla hybrid is a good replacement for rohu to boost the production from the reservoirs.

The morphometric traits of rohu-catla and the reciprocal hybrids have been analysed by different workers in comparison to the parent species. According to Konda Reddy and Varghese (1980 a) the rohu-catla and catla-rohu hybrids exhibit intermediate traits in respect of the morphometric ratios such as length of fish/length of head, length of head/snout length, length of head/width of mouth and length of fish/

length of dorsal fin. In other aspects like length of head/ interorbital space, the hybrids tend more towards catla. Rohu-catla hybrid resembled more of catla in the case of the ratios of length of head/ diametre of eye, while the reciprocal hybrid catla-rohu tends more towards rohu. The taxonomic characters of these hybrids according to these authors with regard to meristic counts have been observed to be similar, (data not shown).

Bhowmick *et al.* (1981) conducted significance test of rohu-catla hybrid for the fin ray counts and nine body proportion ratios and compared them with the parental species rohu and catla. The values showed that the ratio of body length/head width differ significantly from catla ($P<0.01$), whereas standard length/head length and standard length/head width are significant at 5% level ($P<0.05$). Significant differences were also found for fin ray counts of the pectoral fin ($P<0.01$). These authors have also stated that, although the hybrid appears to be intermediate in general appearance to the parent species, the significance analysis clearly indicates that the hybrid has acquired more of maternal (rohu) traits (data not shown). Jana (1993), while working for his Ph.D. made a thorough investigation of the rohu-catla and catla-rohu hybrids with regard to their biological and cytological aspects in comparison to their parental species. According to him the morphometric characters of the hybrids and the parent species indicated that the mean values of trunk length and body weight of the rohu-catla hybrid differed significantly ($P<0.001$). It was also observed that head height and body height in trunk length and head width was significantly different ($P<0.01$) from catla.

Jana (1993) mentioned certain key characteristics of the rohu-catla and catla-rohu hybrids which may help to identify these hybrids without much difficulty. In rohu-catla hybrid (rohu female and catla male) the pectoral fins almost reaches the base of the pelvic fin which is a paternal parent trait in the case of catla, where the pectoral fin almost touches the base of the pelvic fin unlike in rohu which leaves a larger gap between tip of pectoral fin when extended towards the base of the pelvic fin. In the case of the reciprocal hybrid, catla female and rohu male (catla-rohu), the pectoral fin when extended, does not touch the base of the pelvic fin and leaves some gap.

C. Catla × L. Rohita (Catla-rohu) Hybrid

The compatibility in this reciprocal cross (catla-rohu) was moderately higher as indicated by the percentage of fertilization and hatching, 65–80% and 60–80% respectively. The incubation period for the eggs to hatch out was reported to be from 16.40 to 20.50 hrs after fertilization. The size of 72 hrs hatchling varied, from 6.8 to 7.2 mm.

Konda Reddy and Varghese (1983) produced this catla-rohu intergeneric hybrid and studied the embryonic and larval developments and compared with the parental species. They reported that the developmental stages of the hybrid were quite normal and no abnormalities were noticed and were similar to the parental species.

With regard to the external morphological features, both rohu-catla and catla-rohu hybrids were reported to show similar intermediate traits. Catla has an upturned month with infringed lips while rohu has a ventrally pointed mouth with fringed lips. However, both the hybrids possess a terminal mouth with slightly fringed lower lip. The hybrids also possess a pair of maxillary barbels but shorter and not as prominent as in rohu. Konda Reddy and Varghese (1983) and Jana (1993). Konda Reddy and Varghese (1980b) also reported that the mouth parts and gill rakers of both the hybrids are very similar, indicating the same food habits. The observation of Alikunhi *et al.* (1971) have shown the hybrid to grow better than even catla under monoculture as well as when grown with catla. Similarly, Varghese and Sukumaran (1971) also observed the hybrid to grow better than even both the parents. This is in contrary to what was reported by Konda Reddy and Varghese (1980 b). These workers reported that the catla-rohu hybrid (catla female-rohu male) grew very much slower than catla and slightly slower than rohu. Rohu registered 25.2% more than the hybrid while catla registered 108.9% more. These authors concluded that this hybrid is much inferior to catla and slightly slower than rohu.

The reasons for this, according to Konda Reddy and Varghese (1980 b) seem to be that the food habits of these hybrids as reported by Chaudhuri (1973) were basically detritophagic in pond environment and more like rohu. So there appears to be food competition between the hybrids and rohu. Natarajan *et al.* (1976) also reported that in reservoir environment the hybrid was detritophyto-planktophagic and has shown active inclination towards detritus and phytoplankton in preference to zooplankton, which constitutes the chief food of catla.

C. catla × *Cirrhinus Mrigala* (Catla × Mrigal) Hybrid

This intergeneric hybrid cross exhibited relatively high compatibility as evidenced by the high rate of fertilization, over 97%. The fully swollen and developing zygotes measured 5.8 mm while the hatchling measured over 4.0 mm, which is much smaller than the catla-rohu hybrid. Ibrahim (1977) made a thorough investigation of the hybrids produced between *C. catla* female and *Cirrhinus mrigala* male and the reciprocal hybrids and also the hybrid between *L. rohita* female and C. mrigala male. According to him the catla-mrigal hybrid had an elongated body with a terminal but slightly ventrally placed mouth.

The upper lip was slightly thick and the head more like catla. Pectoral, pelvic and anal fins were slightly pinkish but not as much as in mrigal. The hybrid was devoid of barbels. The snout was slightly blunt tending towards the paternal parent, the mrigal. Though not as pronounced as in catla, the hybrid had a slight hump. The body colour was more like mrigal. Diametre of the eye was larger in mrigal than in catla. But in the hybrid the eye was much larger than the parents. Diametre of the eye in relation to head length was 6.89 mm in catla, 5.57 mm in mrigal and 4.129 mm in the hybrid. From the view point of the morphometric characters, the hybrid differed from the parental forms with regard to fin ray counts, dorsal, pectoral and pelvic fin rays, and tended more towards the maternal parent, the catla.

Food and Feeding

Catla, a plankton and surface feeder has 245 gill rakers, which are slender, elongated and suitable for filtering, whereas in mrigal, which is a bottom feeder as also mentioned earlier, has only over 60 short and slender gill rakers which are placed widely. The food niches and food habits of the two parental forms are quite different. The hybrid possesses 91 gill rakers, which is somewhat intermediate but nearer to the paternal parent, the mrigal. Thus the hybrid though gained a little over mrigal in the filtering mechanism, did not gain over the mother. Again, the hybrid with alimentary canal length of 6.15 times the total length, resembles the maternal parent, catla in which it was 6.21 times. However, in mrigal it was much longer, being 18 times, the total length (Ibrahim 1977). Sex ratio of the hybrids was 66.6% females and 33.4% males as reported by Ibrahim (1977).

Growth

Ibrahim (1997) made a detailed study on the growth of catla × mrigal hybrid and the reciprocal hybrids and compared them with the parental species in communal ponds. According to him the catla × mrigal hybrid attained an average growth of 107.7mm/10.3g in 11 months while the parental forms catla and mrigal attained 160.3mm/45.8g and 174.0mm/55.5g respectively. The growth of the hybrid, thus was not as good as the parental forms.

Cirrhinus Mrigala × C. Catla (Mrigal-catla) Hybrid

Unlike in the reciprocal hybrid, this hybrid possessed an up-turned mouth with the upper lip slightly over hanging. Lower lip not fringed and without barbels. The hybrid was reported to resemble the catla (father) with regard to the head and dorsal profile. The hump of catla was also clear in the hybrid. (Ibrahim, 1977).

The hybrid was found to be intermediate to the parents with regard to the anterior portion of the body in relation to its total length. The depth of the body in the hybrid was greater than that of the mrigal as the former has inherited the hump of catla. When compared to catla and rohu, mrigal has a minimum hump. The increase in the body depth due to the hump can be regarded as a positive trait in the hybrid as in the case of the rohu-catla hybrid. The hybrid had an increased size of head in relation to total length (4.58), compared to the mrigal (3.35) and smaller than in catla (3.33). The eye diametre was relatively larger than in catla but not as large as in mrigal. The significant morphometric characters are thus the depth of body and size of head of the hybrid which are positive gains (Ibrahim, 1977).

The fin ray counts in the hybrid with regard to the dorsal and pectoral fins, have shown differences from those of the parents.

The pharyngeal complex in the hybrid with 144 gill rakers has shown significant deviation from both the parents which, as mentioned earlier in mrigal was over 60 and 245 in catla. The increase in the number of gill rakers according to Ibrahim, (1977) is an indication of the hybrid adoption towards plankton food habit as in catla.

Growth

The growth performance of this hybrid (mrigal × catla) according to Ibrahim (1977) was better than either parents. The hybrid grew to an average size of 221.2mm/105g in 11 months compared to growth of catla and mrigal being 160mm/45.8g and 174mm/55.5g respectively.

L.rohita × Cirrhinus Mrigala (Rohu-mrigal) Hybrid

The experiments of Ibrahim (1977) have shown that the hybrid cross had a very high percentage of fertilization (97.9%) indicating the highly compatible nature between the genomes of these two intergeneric species. The incubation period for the eggs to hatch was 15 hours and 30 minutes. The hatchlings, immediately after hatchling measured 3.67mm. By about 92 hours the yolk sack in the hatchling (spawn) was completely absorbed. The hybrid had a ventrally turned mouth with the upper lip over hanging. A maxillary pair of barbels was reported to be present. The lower lip was not fringed as in the case of rohu (maternal parent). The hybrid possessed slightly greenish colour on the dorsal profile as in the case of rohu. However, the ventral side with silvery colour resembled the paternal parent (mrigal). Also, colouration of pectoral, pelvic and caudal fins resembled that of mrigal. The snout was slightly pointed, resembling rohu.

The head length and body height of the hybrid were longer than the father. However, the hybrid showed lower gill-raker length than both the parental. A positive trait in the hybrid was that it possessed a larger number of gill rakers than both the parents. This according to Ibrahim (1977) is a probable indication of the hybrid tending towards consumption of more planktonic food than either of the parents thus providing the hybrid a wider spectrum in feeding. The sex ratio among the hybrids according to Ibrahim (1977) was 35% female, 65% male and some maturing specimens of hybrids were noticed during the third year.

Growth

The growth of this rohu-mrigal hybrid was also good attaining an average size of 242.8mm/143g in 11 months. Among these three intergeneric hybrids, catla-mrigal, mrigal-catla and rohu-mrigal, whose growth performance was evaluated in communal pond along with the parental forms, catla, mrigal and rohu, the mrigal-catla and the rohu-mrigal hybrids have shown a better overall growth performance than the parental forms (Ibrahim, 1977). However, the author mentioned that the growth recorded in this evaluation should not be taken as a complete manifestation of the growth potential as the fishes were confined in a small nursery, which is not considered ideal for maximum growth assessment of fishes. However, the trend observed can be considered significant.

The overall analysis of morphometric and meristic characters of all these hybrids interestingly indicated that the hybrids showed greater divergence from parental forms in characters like body height, head length, eye diametre, gill rakers number and size, and also in the length of the alimentary canal (Ibrahim, 1977).

In the *Cirrhinus mrigala* (mrigal) × *C. catla* (catla), the body height had a gain of 15.8% against mrigal (mother) while it had a loss of 16.5% over catla (father). In head size the hybrid gained 19.9% over mrigal and a loss of 27.6% was observed over catla. In the length of gill rakers the hybrid gained 62.9% over mrigal and there was a loss of 63.3% over catla. Again in eye diametre the hybrid indicated a loss of 1.4% and a gain of 17.6% over the maternal and paternal parental forms respectively.

Considering the other characters, this hybrid has a wider spectrum of feeding in converting more of column-planktonic and bottom feeds. This, possibly could be the reason for a relatively better growth of the hybrid over the parental forms (Ibrahim, 1977).

Basavaraju and Varghese (1980b) studied the growth rates of the *L. rohita* (female) × *Cirrhinus mrigala* (male) and the reciprocal hybrids and compared them with that of the parental species which were also reared along with the hybrids. This study has shown that both the parental species, *L. rohita* and *Cirrhinus mrigala* grew faster than both the hybrids. Again, among the two hybrids the hybrid between *Cirrhinus mrigala* female × *L. rohita* male grew faster than the reciprocal *L. rohita* female × *C. mrigal* male hybrid. Thus, the average weights attained by rohu, mrigal and the reciprocal hybrids, mrigal-rohu and rohu-mrigal were 151.52, 92.37, 72.16 and 61.0g respectively. Whereas the observations of Ibrahim (1977) have shown that growth rate of the rohu (female) × mrigal (male) was better than the parent species.

L. Fimbriatus × C. Catla (Fimbriatus-catla) and C. Catla × L. Fimbriatus (Catla-fimbriatus) Hybrids

Basavaraju *et al.* (1994) studied the morphometric and meristic characters of the hybrid between *L. fimbriatus* female and *C. catla* male in comparison with the parental species. These authors stated that it was difficult to distinguish these reciprocal hybrids.

The fimbriatus-catla hybrid also possessed intermediate characters to parental species in body profile, colouration of fins and structure of lips. Body was deep with a smaller head than in the paternal parent catla. The Mouth was subterminal with a moderately fringed lower lip.

With regard to the morphometric characters, the hybrid resembled catla (male parent) in total length/standard length, length of head/eye diametre and length of head/height of head. Its resemblance with the female parent (fimbriatus) was only in respect of head length/ maximum head width. The intermediate characters were total length/ fork length, total length/pro-dorsal length and total length/least height of caudal peduncle (Basavaraju *et al.*, 1994).

Growth

Evaluation of the hybrid between female *C. catla* and male *L. fimbriatus* (catla-fimbriatus) and the reciprocal hybrid i.e. female *L. fimbriatus* and male *C. catla* (fimbriatus-catla) has been done by Basavaraju *et al.* (1995). These workers reported that the growth of hybrids was much faster than the slow-growing parent, fimbriatus and less than catla when assessed under similar conditions. On conclusion the growth rates recorded were, catla 64.2 fimbriatus 24g, hybrids fimbriatus catla 57.8g and catla-fimbriatus 39.8g.

The growth increment attained by fimbriatus-catla was 153.4 and 45.0% more than fimbriatus and the reciprocal hybrid respectively and not much different from catla. As in the case of the rohu-catla hybrids, these hybrids also exhibit intermediate traits with desirable qualities such as smaller head in the fimbriatus and deeper body as in catla.

The meat yield in the hybrids (62.1% in fimbriatus-catla and 63% in catla-fimbriatus) was higher than in the parents (52.5% in catla and 61.9% in fimbriatus), as in the case of rohu-catla hybrid.

Table: List of intergeneric hybrids between Indian major carps

Female, Male	***Hybrid***	***Comments***
1. *C. catla* × *L.rohita*	(catla-rohu)	Growth rate of hybridsvariable, grow better than both the parents (Verghese and Sukumaran, 1971), grow slower than both the parents (Konda Reddy & Verghese, 1980), fertile hybrids.
2. *L.rohita* × *C. catla*	(rohu-catla)	Hybrids grow faster than rohu. The meat yield in hybrids is higher than in parents. The hybrids are fertile.
3. *C. catla* × *C. mrigala*	(catla-mrigal)	Hybrids grow slower than the parents and are fertile.
4. *C. mrigala* × *C. catla*	(mrigal-catla)	Hybrids grow better than either parents and are fertile.
5. *L.rohita* × *C. mrigala*	(rohu-mrigal)	Growth rate of hybridsvariable, grow better than the parent

Contd...

Female, Male	*Hybrid*	*Comments*
species (Ibrahim, 1977), grow slower than both the parents (Basavaraju and Verghese, 1980b). The hybrids are fertile.		
6. *C. mrigala* × *L.rohita*	(mrigal-rohu)	Hybrids grow slower than both the parents and are fertile.
7. *C. catla* × *L.calbasu*	(catla-kalbasu)	Hybrids grow faster than kalbasu and they have broader feeding spectrum & adaptability.
8. *L. calbasu* × *C. catla*	(kalbasu-catla)	Hybrids exhibited low viability, need further study with regard to survival & performance.
9. *L.calbasu* × *C. mrigala*	(kalbasu-mrigal)	Fertilization in the hybrid cross was normal (80–90%) with 60% hatching rate. The hybrids are fertile.
10. *L.rohita* × *C. reba*	(rohu-reba)	Hybrids exihibit low viability.
11. *L.calbasu* × *C. reba*	(kalbasu-reba)	Hybrids exihibit low viability.
12. *C. catla* × *L. fimbriatus*	(catla-fimbriatus)	Hybrids grow much faster than *L. fimbriatus*. The meat yield is higher than the parents.
13. *L. fimbriatus* × *C. catla*	(fimbriatus-catla)	Hybrids grow much faster than *L. fimbriatus*. The meat yield is higher than the parents.

Intergeneric Hybrids between Indian Major Carps and Common Carp (*Cyprinus Carpio* var. *Communis*)

The hybrids between Indian major carps and common carp (*Cyprinus carpio var. communis*), involving the crosses between common carp male × rohu female, common carp female × males of rohu, catla and mrigal in each cross, respectively have on the whole exhibited normal fertilization and moderate viability, though in certain cases, the crosses involving females of Indian carps with males of common carp, low viability has been exhibited in the hybrids.

Table: *Intergeneric hybrids between Indian major carps and Common carp*

Female, Male	*Hybrid*	*Comments*
1. *L.rohita* × *C. carpio*	(rohu-common carp)	Hybrids exhibited poor survival and are sterile.
2. *C. mrigala* × *C. carpio*	(mrigal-common carp)	Poor survival. The hybrid tends to resemble *C. carpio* and are sterile.

Contd...

Female, Male	*Hybrid*	*Comments*
3. *C. carpio* × *L.rohita*	(common carp-rohu)	Higher survival rate than the reciprocal-cross and dominated by mostly *C. carpio* traits. The hybrids are sterile (aneuploid), grow better under monoculture system.
4. *C. carpio* × *C. catla*	(common carp-catla)	Higher survival rate than the reciprocal-cross and dominated by mostly *C. carpio* traits. The hybrids are sterile (aneuploid), grow better under monoculture system.
5. *C. carpio* × *C. mrigala*	(common carp-mrigal)	Hybrids are inferior to both the parents in growth and are sterile.

Characteristics of Hybrids Produced between Indian Major Carps and Common Carp

The characteristics of these intergeneric hybrids also more or less exhibited intermediate traits, though some traits were characteristic of the maternal parent and in others of the paternal parent.

L. Rohita × C.carpio (Rohu-common Carp) Hybrid

Though the percentage of fertilization appeared to be good, the development in many of the eggs ceased during the incubation period. The bulk of the embryos died during and after hatching and only a few hatchlings could grow up to the adult stage which did not show any signs of maturity.

Morphologically, the hybrid appeared to be intermediate between the parents but statistical analysis of various parameters showed it to be closer to the paternal parent (common carp).

C. Carpio × L. Rohita (Common Carp-rohu) Hybrid

Khan *et al.*, 1990 made a detailed study of this hybrid cross between common carp female and *L.rohita* male. The fertilization rate was as good as in pure common carp crosses. Hatching process was completed within 46.30 hrs at water temperature 27–29.5° C. The dorsally convex and elongated body of the hybrid was broader than that of rohu. In other morphological features the hybrid had a terminal mouth which was also protractile as in common carp. The lips were not fringed but one pair of maxillary and another pair of rostral barbels were present. Regarding the fins, the dorsal fin though long, but was shorter than in common carp. Serrations of the spiny dorsal and anal fin rays which are very conspicuous in common carp, were very feeble in the hybrid. The hybrid resembled the common carp in prepectoral distance, the lateral line position, colour of the body. However, the caudal fin in the hybrid was deeply forked with 24 rays. With regard to size of scales, the hybrid was intermediate. The colour of fins was also intermediate.

The characteristic big belly of common carp was absent in this hybrid, as in the case of mrigal-common carp hybrid, and the shape of the abdomen resembles that of rohu, the paternal parent. The size of the head in the hybrid was intermediate being smaller than that of common carp but bigger than in rohu. On the whole the hybrid had more maternal traits.

Figure: *Intergeneric hybrid-cross between female* C. carpio *and male* L. rohita : a. C. carpio, b. *Hybrid,* c. L. rohita

The survival and growth of common carp-rohu hybrid and common carp under monoculture system were 91.45g and 73% and 76.5g and 71% respectively during a period of 180 days. However, when the hybrid was reared along with the parents, common carp and rohu in equal ratios, the average growth performance and survival were 84.3g and 68.5%, 237g and 70% and 216.7g and 88% respectively. Thus the hybrid has shown slower growth and lower survival than both the parents, when grown together.

Regarding food habits, these hybrids are bottom feeders as the maternal parent and their food items consisted of semidigested crustacean appendages, bottom detritus and sand particles. Diatoms like *Amphora, Navicula* and filamentus alga were also observed in the food. The dress value of the hybrid had a higher percentage of flesh (58%) than that of common carp (47%), but less than in rohu (60.3%). Regarding fertility, the hybrid did not exhibit any signs of maturity after two to four years of age. However, rudiments of testes were observed in some individuals.

C. Carpio × C. Catla (Common Carp-catla) Hybrid

The cross of common carp female with male of *C. catla* showed a relatively high percentage of fertilization, ranging from 65–80%, but in this case also about 35–45% mortality occurred during embryonic development, as in the case of the mrigal-common carp hybrid. Most of the hatchlings were deformed with a curve in the caudal region and died 72 hrs after hatching.

The hybrid possessed a dorsally convex and more elongated body than common carp. A slight hump as in the case of common carp was also present. As in the case of the common carp rohu hybrid, the characteristic belly of the common carp was lacking in this hybrid too. The mouth of the hybrid was terminal and shorter than that of common carp.

Figure: *Intergeneric hybrid-cross between female* C. carpio *and male* C. catla: a. C. carpio, b. *Hybrid (sterile),* c. C. catla

The hybrid attained an average weight of 21.0g in polyculture system and 71g under monoculture system (when reared alone) within a period of two months (Khan *et al.*, 1990).

Thus, over three decades of research on hybridisation among different species of Indian major carps and their evaluation has shown that these major carps though possessing distinct morphological characters, genera, they are highly compatible and able to produce viable and fertile hybrid progenies. Matured interspecific and intergeneric hybrids could be induced to produce F2 progeny or back-cross and triple cross hybrids.

The growth exhibited by most of these hybrids was intermediate i.e. better than the slow growing parent as in the case of the interspecific hybrid between *L. rohita* and *L. calbasu* or the intergeneric hybrids between *L.rohita* and *C. catla* or *C. catla* and *Cirrhinus mrigala* etc.

The gain in the growth in these hybrids over their slow growing parents appears to be due to the inheritance of this trait from the fast growing parent, rohu in the case of *L. rohita-L. calbasu* hybrid and catla in the case of *L.rohita-C. catla* hybrid and so on. Chevassus (1983) also expressed the opinion that the intermediate nature of the growth in the hybrids, may be due to the partial transmission of this trait to the hybrid by the fast growing parent.

The growth of some hybrids like the rohu-catla or mrigal-catla or with their reciprocal cross appears to be due to the available food in the medium and also could be due to the variations that probably existed in the rearing medium under different agroclimatic zones.

With regard to the hybrid crosses between Indian major carps and Chinese grass carp, silver carp and big head carp, no compatibility was observed in any of the hybrid progenies as almost all of them died either during the embryonic development or soon after hatching, maximum within a week, except some stray hybrids produced between *C. catla* and *H. molitrix* (silver carp).

Again, relatively better survival of the hybrid progeny was observed in the crosses between female common carp × males of Indian major carps, viz., *C. catla, L. rohita* and *Cirrhinus mrigala* compared to the reciprocal hybrid crosses particularly between female *L. rohita* and male common carp. These reciprocal hybrids followed the maternal parent in most of the embryonic developmental events particularly, duration of incubation, commencement of cleavage etc. However, the reason the viability of hybrid progeny was poor in crosses where Indian major carps were the maternal parents appears to be due to the haploid nature of most of those hybrid progenies as indicated by their curved caudal region probably caused due to the relatively larger size of the head of the penetrating sperm of common carp (with n=50–51 chromosomes) through the smaller micropyle of the major carp egg (with n=25 chromosomes). Thus often preventing the genome input from the male and thereby proper fertilization.

However, further investigation may be necessary, especially on hybrids between Indian major carps and minor carps and between Indian major carps and common carp.

It is essential to prevent indiscriminate hybridisation by checking the mixed spawning of Indian carps. These carps, as has been observed, are highly compatible to interbreed. Also, since almost all interspecific and intergeneric hybrids were found fertile, we should be all the more cautious to check their cross-breeding which may cause serious damage to the original gene pool through genetic introgression. (Padhi and Mandal, 1994, Mishra *et al.*, 1998).

Some of these hybrids may be very useful for stocking reservoirs in India, where they are observed to thrive well and contribute substantially to the production (Natarajan, *et al.*, 1976).

Evaluation of Meat of Hybrids

Basavaraju and Varghese (1980) have made some study on organoleptic evaluation of the meat of rohu-mrigal and mrigal-rohu and compared it to that of the parental species, rohu and mrigal. The test was done for both raw and cooked meat. The attributes tested were colour gloss of skin, odour of flesh, colour of flesh, and texture for raw fish and colour, texture, taste and flavour for cooked meat. Analysis of variance (Krammer and Twigg, 1970, Basavaraju and Varghese 1980) showed that the uncooked flesh of hybrids was found to be inferior in quality to that of the parental species, while no significant difference was found with regard to the cooked flesh.

Mixed Spawning and Hybridisation

After the development of induced breeding technique for Indian major carps, production of pure seed of these carps became possible. Dependence of the fish culturist for their seed from riverine sources gradually decreased for many reasons. After the advent of the intensive carp culture system, more and more farmers started taking up fish culture. Consequently there has been a growing demand for the seed of major carps. Based on the demand of fish farmers, the seed producers that used Chinese-type hatcheries produced the seed of all three major carps, catla, rohu and mrigal, by breeding them together in a common bedding pool through the administration of hormones.

This practice is mostly followed in the state of West Bengal where 70% of the carp seed requirement of the country is met (Padhi and Mandal, 1994). This kind of seed production is generally resorted to by the seed producer, not only to supply seed of all the three species simultaneously in the required quantities, but also to save time as the breeding span of these carps lasts usually for only one and a half to two months.

Thus the seed producer's aim is to supply the seed as per the demand and earn as much money as possible, without realising that there may be genetic consequences of the practice.

Mishra. *et al.* (1998) have assessed the percentage of hybridisation taking place in such mixed spawning practices. According to their observations the rate of what they termed as spontaneous/inadvertent hybridisation (intergeneric) ranged from 7.24 to 9.24% in a fish seed hatchery, they studied in the state of Uttar Pradesh. The hybrids were identified as catla-mrigal and, rohu-mrigal.

The growth performance of these hybrids as observed by Mishra, *et al.* (1998) was that the hybrids in general were found to possess significantly higher lengths and weights than their parent species. The growth of catla-rohu was found to be highest among all. The rohu-mrigal hybrids performed better than their parents. Performance of catla-mrigal hybrids was better than mrigal, but not catla.

Though all the three species bred together (mixed spawning), the incidence of hybridisation (7.24 to 9.24%) is not relatively very high in spite of the compatibility of these species to hybridise. The probable reasons for this could be that Indian carps exhibit courting behaviour while spawning where the female is usually followed (chased) by several, males very closely.

Here, males might be following their like species during the chase, ultimately resulting in the release of eggs by the female and simultaneously release of milt by the males.

It was also reported that the male coils round the abdomen of the female and releases milt when she releases the eggs The spontaneous/inadvertent hybridisation which is reported to take place during mixed spawn may be due to accidental fertilization of the eggs of one species by the sperm of another species which might have been released in close proximity.

Species specific courtship behaviour and generally species recognition keep the accidental level of hybridisation low. However, the presence of even a few number of hybrids among the lot of other pure species (Indian major carps) may be detrimental to the genetic quality in the long run.

4

Aquaculture of Salmon

Salmon, along with carp, are the two most important fish groups in aquaculture. In 2007, the aquaculture of salmon and salmon trout was worth US$10.7 billion. The most commonly farmed salmon is the Atlantic salmon. Other commonly farmed fish groups include tilapia, catfish, sea bass, bream and trout.

Figure: *Salmon farm in the archipelago of Finland.*

Salmon aquaculture production grew over ten-fold during the 25 years from 1982 to 2007. Leading producers of farmed salmon are Norway with 33 percent, Chile with 31 percent, and other European producers with 19 percent. There is currently much controversy about the ecological and health impacts of intensive salmon aquaculture. There are particular concerns about the impacts on wild salmon and other marine life. Some of this controversy is part of a major commercial competitive fight for market share and price between Alaska commercial salmon fishermen and the rapidly evolving salmon aquaculture industry.

Methods

The aquaculture of salmon is the farming and harvesting of salmon under controlled conditions. Farmed salmon can be contrasted with wild salmon captured using commercial fishing techniques. However, the concept of "wild" salmon as used by the Alaska Seafood Marketing Institute includes stock enhancement fish produced in hatcheries that have historically been considered ocean ranching. The percentage of the Alaska salmon harvest resulting from ocean ranching depends upon the species of salmon and location, however it is all marketed as "wild Alaska salmon".

***Figure:** A salmon farm which holds yearlings for up to two years. Many hold broodstock for even longer in these conditions to help ensure large, sexually mature adults.*

Methods of salmon aquaculture originated in late 18th century fertilization trials in Europe. In the late 19th century, salmon hatcheries were used in Europe and North America. From the late 1950s, enhancement programs based on hatcheries were established in the United States, Canada, Japan and the USSR. The contemporary technique using floating sea cages originated in Norway in the late 1960s.

Salmon are usually farmed in 2 stages and in some places maybe more. First, the salmon are hatched from eggs and raised on land in freshwater tanks. When they are 12 to 18 months old, the smolt (juvenile salmon) are transferred to floating sea cages or net pens anchored in sheltered bays or fjords along a coast. This farming in a marine environment is known as mariculture. There they are fed pelleted feed for another 12 to 24 months, when they are harvested.

Figure: *Very young fertilised salmon eggs; notice the developing eyes and vertebral column.*

Norway produces 33 percent of the world's farmed salmon, and Chile produces 31 percent. The coastlines of these countries have suitable water temperatures and many areas well protected from storms. Chile is close to large forage fisheries which supply fish meal for salmon aquaculture. Scotland and Canada are also significant producers.

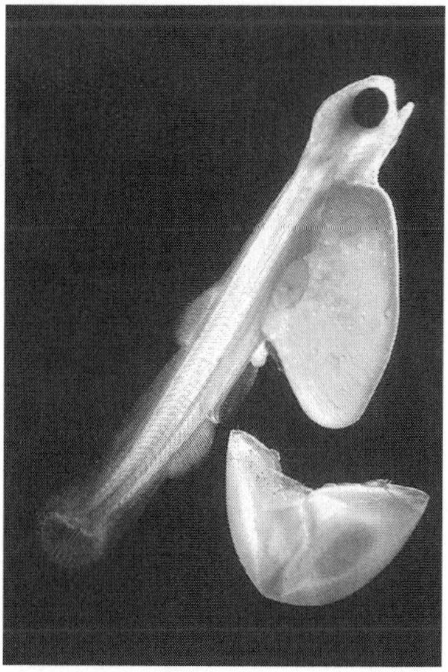

Figure: *Salmon egg hatching. In about 24hrs it will be a fry without the yolk sac.*

Modern salmon farming systems are intensive. Their ownership is often under the control of huge agribusiness corporations, operating mechanised assembly lines on an industrial scale. In 2003, nearly half of the world's farmed salmon was produced by just five companies.

Hatcheries

Modern commercial hatcheries for supplying salmon smolts to aquaculture net pens have been shifting to Recycle Aquaculture Systems (RAS) where the water is recycled within the hatchery. This allows location of the hatchery to be independent of a significant fresh water supply and allows economical temperature control to both speed up and slow down the growth rate to match the needs of the net pens.

Conventional hatchery systems operate flow through where spring water or other water source flow into the hatchery. The eggs are then hatched in trays and the salmon smolts produced in raceways. The waste products from the growing salmon fry and the feed are usually discharged into the local river. Conventional flow through hatcheries, such as are being used by most of the Alaska ocean ranching enhancement hatcheries use more than 100 tons of water to produce to produce a kg of smolts.

An alternative method to hatching in freshwater tanks is to use spawning channels. These are artificial streams, usually parallel to an existing stream with concrete or rip-rap sides and gravel bottoms. Water from the adjacent stream is piped into the top of the channel, sometimes via a header pond to settle out sediment.

Spawning success is often much better in channels than in adjacent streams due to the control of floods which in some years can wash out the natural redds. Because of the lack of floods, spawning channels must sometimes be cleaned out to remove accumulated sediment.

The same floods which destroy natural redds also clean them out. Spawning channels preserve the natural selection of natural streams as there is no temptation, as in hatcheries, to use prophylactic chemicals to control diseases.

However, exposing fish to wild parasites and pathogens using uncontrolled water supplies, combined with the high cost of spawning channels, makes this technology unsuitable for salmon aquaculture businesses. This type of technology is only useful for stock enhancement programs.

Sea Cages

Sea cages, also called sea pens or net pens, are usually made of mesh framed with steel or plastic. They can be square or circular, 10 to 32 metres across and 10 metres deep, with volumes between 1,000 and 10,000 cubic metres. A large sea cage can house up to 90,000 fish.

They are usually placed side by side to form a system called a *seafarm* or *seasite*, with a floating wharf and walkways along the net boundaries. Additional nets can also surround the seafarm to keep out predatory marine mammals. Stocking densities range from 8 to 18 kilograms per cubic metre for Atlantic salmon and 5 to 10 kilograms per cubic metre for Chinook salmon.

In contrast to closed or recirculating systems, the open net cages of salmon farming lower production costs, but provide no effective barrier to the discharge of wastes, parasites and disease into the surrounding coastal waters. Farmed salmon in open net cages can escape into wild habitats, for example, during storms.

An emerging wave in aquaculture is applying the same farming methods used for salmon to other carnivorous finfish species, such as cod, bluefin tuna, halibut and snapper. However, this is likely to have the same environmental drawbacks as salmon farming.

A second emerging wave in aquaculture is the development of copper alloys as netting materials. Copper alloys have become important netting materials because they are antimicrobial (i.e., they destroy bacteria, viruses, fungi, algae, and other microbes) and they therefore prevent biofouling (i.e., the undesirable accumulation, adhesion, and growth of microorganisms, plants, algae, tubeworms, barnacles, mollusks, and other organisms). By inhibiting microbial growth, copper alloy aquaculture cages avoid costly net changes that are necessary with other materials. The resistance of organism growth on copper alloy nets also provides a cleaner and healthier environment for farmed fish to grow and thrive.

Feeding

Salmon are carnivorous and are currently being fed compound fish feeds containing fish meal and other feed ingredients, ranging from wheat byproducts to soybean meal and feather meal. Being an aquatic carnivore, salmon don't tolerate or properly metabolise many plant based carbohydrates and use fats instead of carbohydrates as a primary energy source. With the amount of world wide fish meal production being almost a constant amount for the last 30+ years and at maximum sustainable yield (MSY), much of the fish meal market has shifted from chicken and pig feed to fish and shrimp feeds as aquaculture has grown in this time period. Work continues on substituting vegetable proteins and protein concentrates for fish meal in the salmon diet.

Many substitutions for fish meal are known and diets containing zero fish meal are possible. However, commercial economic animal diets are determined by least cost linear programming models that are effectively competing with similar models for chicken and pig feeds for the same feed ingredients and these models show that fish meal is more useful in aquatic diets than in chicken diets, where they can make the chickens taste like fish. Unfortunately, this substitution can result in lower levels of the highly valued omega-3 content in the farmed product. However, when vegetable oil is used in the growing diet as an energy source and a different finishing diet containing high omega-3 content fatty acids from either fish oil, algae oils or some vegetable oils are used a few months before harvest, this problem is eliminated. At the present time, more than 50 percent of the world fish oil production is feed to farmed salmon. Farm raised salmon are also fed the carotenoids astaxanthin and canthaxanthin, so that their flesh colour matches wild salmon, which also contain the same carotenoid pigments from their diet in the wild. On a dry-dry basis, it takes 2–4 kg of wild caught fish to produce one kg of salmon. Wild salmon require about 10kg of forage fish to produce a kg of salmon, as part of the normal trophic level energy transfer. The difference between the two numbers is related to farmed salmon feed containing other ingredients beyond fish meal and the fact that farmed fish don't spend a lot of metabolic energy catching a dinner that doesn't want to be caught.

Harvesting

Modern harvesting methods are shifting towards using wet well ships to transport the live salmon to the processing plant. This allows the fish to be killed, bled, and filleted before rigor has occurred. This results in superior product quality to the customer along with more humane processing. To obtain maximum quality, it is necessary to minimise the level of stress in the live salmon until actually being electrically and percussively killed and the gills slit for bleeding. These improvements in processing time and freshness to the final customer are commercially significant and forcing the commercial wild fisheries to upgrade their processing to the benefit of all seafood consumers.

An older method of harvesting is to use a sweep net, which operates a bit like a purse seine net. The sweep net is a big net with weights along the bottom edge. It is stretched across the pen with the bottom edge extending to the bottom of the pen. Lines attached to the bottom corners are raised, herding some fish into the purse, where they are netted.

Before killing, the fish are usually rendered unconscious in water saturated in carbon dioxide, although this practice is being phased out in some countries due to ethical and product quality concerns. More advanced systems use a percussive-stun harvest system that kills the fish instantly and humanely with a blow to the head from a pneumatic piston. They are then bled by cutting the gill arches and immediately immersing them in iced water. Harvesting and killing methods are designed to minimise scale loss, and avoid the fish releasing stress hormones, which negatively affect flesh quality.

Wild Versus Farmed

Wild salmon are captured from wild habitats using commercial fishing techniques. Most wild salmon is caught in North American, Japanese and Russian fisheries. The following table shows the changes in production of wild salmon and farmed salmon over a period of 25 years, as reported by the FAO. Russia, Japan and Alaska all operate major hatchery based stock enhancement programs that are really ocean ranching. The resulting fish hatchery fish are defined as "wild" for FAO and marketing purposes.

Table: Salmon production in tonnes by species

	1982		*2007*	
Species	***Wild***	***Farmed***	***Wild***	***Farmed***
Atlantic salmon	10,326	13,265	2,989	1,433,708
Salmon trout		171,946		604,695
Coho salmon	42,281	2,921	17,200	115,376
Chinook salmon	25,147		8,906	11,542
Pink salmon	170,373		495,986	
Chum salmon	182,561		303,205	
Sockeye salmon	128,176		164,222	

Table: Total salmon production

	1982		*2007*	
	tonnes	***percent***	***tonnes***	***percent***
Wild	558,864	75	992,508	31
Farmed	188,132	25	2,165,321	69
Overall	746,996		3,157,831	

Issues

There is currently much controversy about the ecological and health impacts of intensive salmon aquaculture. There are particular concerns about the impacts on wild salmon and other marine life and on the incomes of commercial salmon fishermen. The interest of the commercial fishermen are trying to prevent competition.

Disease and Parasites

In 1972, Gyrodactylus, a monogenean parasite, spread from Norwegian hatcheries to wild salmon, and devastated some wild salmon populations.

In 1984, infectious salmon anemia (ISAv) was discovered in Norway in an Atlantic salmon hatchery. Eighty percent of the fish in the outbreak died. ISAv, a viral disease, is now a major threat to the viability of Atlantic salmon farming. It is now the first of the diseases classified on List One of the European Commission's fish health regime. Amongst other measures, this requires the total eradication of the entire fish stock should an outbreak of the disease be confirmed on any farm.

ISAv seriously affects salmon farms in Chile, Norway, Scotland and Canada, causing major economic losses to infected farms. As the name implies, it causes severe anemia of infected fish. Unlike mammals, the red blood cells of fish have DNA, and can become infected with viruses. The fish develop pale gills, and may swim close to the water surface, gulping for air. However, the disease can also develop without the fish showing any external signs of illness, the fish maintain a normal appetite, and then they suddenly die. The disease can progress slowly throughout an infected farm and, in the worst cases, death rates may approach 100 percent. It is also a threat to the dwindling stocks of wild salmon. Management strategies include developing a vaccine and improving genetic resistance to the disease.

In the wild, diseases and parasites are normally at low levels, and kept in check by natural predation on weakened individuals. In crowded net pens they can become epidemics. Diseases and parasites also transfer from farmed to wild salmon populations. A recent study in British Columbia links the spread of parasitic sea lice from river salmon farms to wild pink salmon in the same river. The European Commission (2002) concluded "The reduction of wild salmonid abundance is also linked to other factors but there is more and more scientific evidence establishing a direct link between the number of lice-infested wild fish and the presence of cages in the same estuary." It is reported that wild salmon on the west coast of Canada are being driven to extinction by sea lice from nearby salmon farms. These predictions have been disputed by other scientists and recent harvests have indicated that the predictions were in error. Antibiotics and pesticides are often used to control the diseases and parasites.

Pollution and Toxins

Salmon farms are typically sited in marine ecosystems with good water quality, high water exchange rates, current speeds fast enough to prevent pollution of the bottom but slow enough to prevent pen damage, protection from major storms, reasonable water depth and a reasonable distance from major infrastructure such as ports, processing plants and logistical facilities like airports. Logistical considerations are significant and feed and maintenance labour must be transported to the facility and the product returned. Siting decisions are complicated by complex politically driven permitting problems in many countries that prevents optimal locations for the farms.

In sites without adequate currents there can be an accumulation of heavy metals on the benthos (seafloor) near the salmon farms, particularly copper and zinc.

Toxins are also found in the flesh of farmed salmon. A 2004 study, reported in *Science*, analysed farmed and wild salmon for organochlorine contaminants. They found the contaminants were higher in farmed salmon. Within the farmed salmon, European (particularly Scottish) salmon had the highest levels, and Chilean salmon the lowest. A follow up study confirmed this, and found levels of dioxins, chlorinated pesticides, PCBs and other toxins up to ten times greater in farmed salmon than wild Pacific salmon. On a positive note, further research using the same fish samples used in the previous study, showed that farmed salmon contained levels of beneficial fatty acids that were two to three times higher than wild salmon. A follow up benefit-risk analysis on salmon consumption balanced the cancer risks with the (n-3) fatty acid advantages of salmon consumption. They found that recommended levels of (n-3) fatty acid consumption can be achieved eating farmed salmon with acceptable carcinogenic risks, but recommended levels of EPA+DHA intake cannot be achieved solely from farmed (or wild) salmon without unacceptable carcinogenic risks. The conclusions were that

> *"...consumers should not eat farmed fish from Scotland, Norway and eastern Canada more than three times a year; farmed fish from Maine, western Canada and Washington state no more than three to six times a year; and farmed fish from Chile no more than about six times a year. Wild chum salmon can be consumed safely as often as once a week, pink salmon, Sockeye and Coho about twice a month and Chinook just under once a month."*

Impact on Wild Salmon

Farmed salmon can, and often do, escape from sea cages. If the farmed salmon is not native, it can compete with native wild species for food and habitat. If the farmed salmon is native, it can interbreed with the wild native salmon.

Such interbreeding can reduce genetic diversity, disease resistance and adaptability. In 2004, about 500,000 salmon and trout escaped from ocean net pens off Norway. Around Scotland, 600,000 salmon were released during storms. Commercial fishermen targeting wild salmon not infrequently catch escaped farm salmon. At one stage, in the Faroe Islands, 20 to 40 percent of all fish caught were escaped farm salmon. Sea lice, particularly *Lepeophtheirus salmonis* and various *Caligus* species, including *Caligus clemensi* and *Caligus rogercresseyi*, can cause deadly infestations of both farm-grown and wild salmon. Sea lice are ectoparasites which feed on mucous, blood, and skin, and migrate and latch onto the skin of wild salmon during free-swimming, planktonic *nauplii* and *copepodid* larval stages, which can persist for several days. Large numbers of highly populated, open-net salmon farms can create exceptionally large concentrations of sea lice; when exposed in river estuaries containing large numbers of open-net farms, many young wild salmon are infected, and do not survive as a result. Adult salmon may survive otherwise critical numbers of sea lice, but small, thin-skinned juvenile salmon migrating to sea are highly vulnerable. On the Pacific coast of Canada, the louse-induced mortality of pink salmon in some regions is commonly over 80%.

A 2008 meta-analysis of available data shows that salmon farming reduces the survival of associated wild salmon populations. This relationship has been shown to hold for Atlantic, steelhead, pink, chum, and coho salmon. The decrease in survival or abundance often exceeds 50 percent. However, these studies are all correlation analysis and correlation doesn't equal causation, especially when similar salmon declines were occurring in Oregon and California, which have no salmon aquaculture or marine net pens. Independent of the predictions of the failure of salmon runs in Canada indicated by these studies, the wild salmon run in 2010 was a record harvest.

Wild salmon are anadromous. They spawn inland in fresh water and when young migrate to the ocean where they grow up. Most salmon return to the river where they were born, although some stray to other rivers. There is concern about of the role of genetic diversity within salmon runs. The resilience of the population depends on some fish being able to survive environmental shocks, such as unusual temperature extremes. It is also unclear what the effect of hatchery production has been on the genetic diversity of salmon.

Genetic Modification

Salmon have been genetically modified in laboratories so they can grow faster. There is opposition to the commercial use of these fish, and, so far, no approval has been given. A Canadian company, Aqua Bounty Farms, has developed a modified Atlantic salmon which grows several times faster, and is more disease resistant and cold tolerant. This was achieved using a chinook salmon gene sequence affecting growth hormones, and a promoter sequence from the ocean pout affecting antifreeze production. Normally, salmon produce growth hormones only in the presence of light. The modified salmon doesn't switch growth hormone production off. The company first submitted the salmon for FDA approval in 1996. A concern with transgenic salmon is what might happen if they escape into the wild. One study, in a laboratory setting, found that modified salmon mixed with their wild cohorts were aggressive in competing, but ultimately failed.

Impact on Forage Fish

The use of forage fish for fish meal production has been almost a constant for the last thirty years and at the maximum sustainable yield, while the market for fish meal has shifted from chicken, pig and pet food to aquaculture diets. The fact that this market shift at constant production is an economic decision having no impact on the forage fish harvest rates for fish meal implies that the development of salmon aquaculture had no impact on forage fish harvest rates.

Fish do not actually produce omega-3 fatty acids, but instead accumulate them from either consuming microalgae that produce these fatty acids, as is the case with forage fish like herring and sardines, or, as is the case with fatty predatory fish, like salmon, by eating prey fish that have accumulated omega-3 fatty acids from microalgae. To satisfy this requirement, more than 50 percent of the world fish oil production is fed to farmed salmon.

In addition, salmon require nutritional intakes of protein, protein which is often supplied to them in the form of fish meal as the lowest cost alternative protein. Consequently, farmed salmon consume more fish than they generate as a final product. To produce one pound of farmed salmon, products from several pounds of wild fish are fed to them. As the salmon farming industry expands, it requires a higher percentage of the wild forage fish for feed, at a time when seventy five percent of the worlds monitored fisheries are already near to or have exceeded their maximum sustainable yield.

Salmon Aquaculture Dialogue

In 2004 the World Wide Fund for Nature (WWF) initiated the *Salmon Aquaculture Dialogue*. The aim of the dialogue is to produce an environmental standard for farmed salmon by 2010. The WWF have identified what they call "seven key environmental and social impacts", which they characterise as follows:

1. Benthic impacts and siting: Chemicals and excess nutrients from food and feces associated with salmon farms can disturb the flora and fauna on the ocean bottom (benthos).
2. Chemical inputs: Excessive use of chemicals - such as antibiotics, anti-foulants and pesticides - or the use of banned chemicals can have unintended consequences for marine organisms and human health.

3. Disease/parasites: Viruses and parasites can transfer between farmed and wild fish, as well as among farms.
4. Escapes: Escaped farmed salmon can compete with wild fish and interbreed with local wild stocks of the same population, altering the overall pool of genetic diversity.
5. Feed: A growing salmon farming business must control and reduce its dependency upon fishmeal and fishoil - a primary ingredient in salmon feed - so as not to put additional pressure on the world's fisheries. Fish caught to make fishmeal and oil currently represent one-third of the global fish harvest.
6. Nutrient loading and carrying capacity: Excess food and fish waste in the water have the potential to increase the levels of nutrients in the water. This can cause the growth of algae, which consumes oxygen that is meant for other plant and animal life.
7. Social issues: Salmon farming often employs a large number of workers on farms and in processing plants, potentially placing labour practices and worker rights under public scrutiny. Additionally, conflicts can arise among users of the shared coastal environment.

Hatch and Release

Another form of salmon production, which is safer but less controllable, is to raise salmon in hatcheries until they are old enough to become independent. They are then released into rivers, often in an attempt to increase the salmon population. This practice was very common in countries like Sweden before the Norwegians developed salmon farming, but is seldom done by private companies, as anyone may catch the salmon when they return to spawn, limiting a company's chances of benefiting financially from their investment. Because of this, the method has mainly been used by various public authorities and non profit groups like the Cook Inlet Aquaculture Association as a way of artificially increasing salmon populations in situations where they have declined due to overharvest, construction of dams, and habitat destruction or disruption. Unfortunately, there can be negative consequences to this sort of population manipulation, including genetic "dilution" of the wild stocks, and many jurisdictions are now beginning to discourage supplemental fish planting in favour of harvest controls and habitat improvement and protection. A variant method of fish stocking, called ocean ranching, is under development in Alaska. There, the young salmon are released into the ocean far from any wild salmon streams. When it is time for them to spawn, they return to where they were released where fishermen can then catch them.

Species

Atlantic Salmon

Figure: *Atlantic salmon*

In their natal streams, Atlantic salmon are considered a prized recreational fish, pursued by avid fly anglers during its annual runs. At one time, the species supported an important commercial fishery and a supplemental food fishery. However, the wild Atlantic salmon fishery is commercially dead; after extensive habitat damage and overfishing, wild fish make up only 0.5% of the Atlantic salmon available in world fish markets. The rest are farmed, predominantly from aquaculture in Chile, Canada, Norway, Russia, the UK and Tasmania in Australia.

Atlantic salmon is, by far, the species most often chosen for farming. It is easy to handle, it grows well in sea cages, commands a high market value and it adapts well to being farmed away from its native habitats.

Adult male and female fish are anaesthetised. Eggs and sperm are "stripped", after the fish are cleaned and cloth dried. Sperm and eggs are mixed, washed, and placed into fresh water. Adults recover in flowing, clean, well aerated water. Some researchers have studied cryopreservation of the eggs.

Fry are generally reared in large freshwater tanks for 12 to 20 months. Once the fish have reached the smolt phase, they are taken out to sea where they are held for up to two years. During this time the fish grow and mature in large cages off the coasts of Canada, the USA, or parts of Europe.

Generally, cages are made of two nets. Inner nets, which wrap around the cages, hold the salmon. Outer nets, which are held by floats, keep predators out. Many Atlantic salmon escape from cages at sea. Those salmon who further breed tend to lessen the genetic diversity of the species leading to lower survival rates, and lower catch rates. On the West Coast of Northern America, the non-native salmon can be an invasive threat, especially in Alaska and parts of Canada. This causes them to compete with native salmon for resources. Extensive efforts are underway to prevent escapes and the spread of Atlantic salmon in the Pacific and elsewhere.

Atlantic salmon which have escaped from the aquaculture industry have also been found breeding in rivers tributary to the Pacific Ocean in British Columbia on Canada's west coast.

Worldwide, in 2007, 1,433,708 tonnes of Atlantic salmon were harvested with a value of 7.578 billion US dollars.

Salmon Trout

Figure: *Rainbow trout*

Figure: *Male ocean phase steelhead salmon*

In 1992 the salmon trout, *Salmo gairdneri*, was reclassified as a true salmon, *Oncorhynchus mykiss*. Salmon trout take an anadromous form and a non-anadromous form. The anadromous form, who migrate between lakes and rivers and the ocean, are known as steelhead salmon or ocean trout, while the non-anadromous form, who spend their entire life in freshwater, are known as rainbow trout.

Salmon trout are raised in many countries throughout the world. Since the 1950s production has grown exponentially, particularly in Europe and recently in Chile. Worldwide, in 2007, 604,695 tonnes of farmed salmon trout were harvested with a value of 2.589 billion US dollars. The largest producer is Chile. In Chile and Norway, the ocean cage production of steelhead has expanded to supply export markets. Inland production of rainbow trout to supply domestic markets has increased strongly in countries such as Italy, France, Germany, Denmark and Spain. Other significant producing countries include the USA, Iran, Germany and the UK.

Salmon trout have tender flesh and a mild, somewhat nutty flavour. Steelhead meat is pink like that of other salmon, and is more flavourful than the light-coloured meat of rainbow trout. Both are highly desired food.

However, farmed trout and those taken from certain lakes have a pronounced earthy flavour which many people find unappealing; many shoppers therefore make it a point to ascertain the source of the fish before buying. Salmon trout that are wild have a diet of scuds (freshwater shrimp), insects such as flies, and crayfish are the most appealing. Dark red/orange meat indicates that it is either an anadromous steelhead or a farmed rainbow trout given a supplemental diet with a high iodine content. The resulting pink flesh is marketed under monikers like Ruby Red or Carolina Red.

The salmon trout is especially susceptible to enteric redmouth disease. There has been considerable research conducted on redmouth disease, as its implications for salmon trout farmers are significant. The disease does not affect humans.

Coho Salmon

Figure: *Male ocean phase Coho salmon*

Coho salmon mature after only one year in the sea, so two separate broodstocks (spawners) are needed, alternating each year. Broodfish are selected from the salmon in the seasites and transferred to freshwater tanks for maturation and spawning."

Worldwide, in 2007, 115,376 tonnes of farmed Coho salmon were harvested with a value of 456 million US dollars. Chile, with about 90 percent of word production, is the primary producer with Japan and Canada producing the rest.

Chinook Salmon

Figure: *Male ocean phase Chinook*

Figure: *Male freshwater phase Chinook*

In Alaska, Chinook salmon are the state fish, and are known as "king salmon" because of their large size and flavourful flesh. Those from the Copper River in Alaska are particularly known for their colour, rich flavour, firm texture, and high Omega-3 oil content.

Worldwide, in 2007, 11,542 tonnes of farmed Chinook salmon were harvested with a value of 83 million US dollars. New Zealand is the largest producer of farmed king salmon, accounting for over half of world production (7,400 tonnes in 2005). Most of the salmon are farmed in the sea (mariculture) using a method sometimes called sea-cage ranching. Sea-cage ranching takes place in large floating net cages, about 25 metres across and 15 metres deep, moored to the sea floor in clean, fast-flowing coastal waters. Smolt (young fish) from freshwater hatcheries are transferred to cages containing several thousand salmon, and remain there for the rest of their life. They are fed fishmeal pellets high in protein and oil.

King salmon are also farmed in net cages placed in freshwater rivers or raceways, using techniques similar to those used for sea-farmed salmon. A unique form of freshwater salmon farming occurs in some hydroelectric canals. A site in Tekapo, fed by fast cold waters from the Southern Alps, is the highest salmon farm in the world, 677 metres above sea level.

Before they are killed, cage salmon are anaesthetised with a herbal extract. They are then spiked in the brain. The heart beats for a time as the animal is bled from its sliced gills. This method of relaxing the salmon when it is killed produces firm, long-keeping flesh. Lack of disease in wild populations and low stocking densities used in the cages means that New Zealand salmon farmers do not use antibiotics and chemicals that are often needed elsewhere.

Timeline

- 1527: The life history of the Atlantic salmon is described by Hector Boece of the University of Aberdeen, Scotland.
- 1763: Fertilization trials for Atlantic salmon take place in Germany. Later biologists refined these in Scotland and France.
- 1854: Salmon spawing beds and rearing ponds built along the bank of a river by the Dohulla Fishery, Ballyconneely, Ireland.
- Late 19th century: Salmon hatcheries are used in Europe, North America, and Japan to enhance wild populations.
- Late 1960s: First salmon farms established in Norway and Scotland.
- Early 1970s: Salmon farms established in North America.
- 1975: Gyrodactylus, a small monogenean parasite, spreads from Norwegian hatcheries to wild salmon, probably by means of fishing gear, and devastates some wild salmon populations.
- Late 1970s: Salmon farms established in Chile and New Zealand.
- 1984: Infectious salmon anemia, a viral disease, is discovered in a Norwegian salmon hatchery. Eighty percent of the involved fish die.
- 1985: Salmon farms established in Australia.
- 1987: First reports of escaped Atlantic salmon being caught in wild Pacific salmon fisheries.
- 1988: A storm hits the Faroes Islands releasing millions of Atlantic salmon.
- 1989: Furunculosis, a bacterial disease, spreads through Norwegian salmon farms and wild salmon.
- Early 1990s: Sea lice from Atlantic salmon farms infest wild sea trout around Ireland, resulting in the collapse of their fisheries.
- 1996: World farmed salmon production exceeds wild salmon harvest.

- 2007: A 10-square-mile (26 km^2) swarm of Pelagia noctiluca jellyfish wipes out a 100,000 fish salmon farm in Northern Ireland.

Freshwater Fish

Figure: *Tench are common freshwater fish all over the world.*

Freshwater fish are fish that spend some or all of their lives in fresh water, such as rivers and lakes, with a salinity of less than 0.05%. These environments differ from marine conditions in many ways, the most obvious being the difference in levels of salinity. To survive fresh water, the fish need a range of physiological adaptations in order to keep the ion concentration of their bodies balanced.

41% of all known species of fish are found in fresh water. This is primarily due to the rapid speciation that the scattered habitats make possible. When dealing with ponds and lakes, one might use the same basic models of speciation as when studying island biogeography.

Physiology

Fresh water fish differ physiologically from salt water fish in several respects. Their gills must be able to diffuse dissolved gasses while keeping the salts in the body fluids inside. Their scales reduce water diffusion through the skin: freshwater fish that have lost too many scales will die. They also have well developed kidneys to reclaim salts from body fluids before excretion.

Migrating Fish

Many species of fish do reproduce in freshwater, but spend most of their adult lives in the sea. These are known as anadromous fish, and include, for instance, salmon, trout and three-spined stickleback. Some other kinds of fish are, on the contrary, born in salt water, but live most of or parts of their adult lives in fresh water; for instance the eels. These are known as catadromous fish.

Figure : *Sturgeons are found both in anadromous and fresh water stationary forms*

Species migrating between marine and fresh waters need adaptations for both environments; when in salt water they need to keep the bodily salt concentration on a level lower than the surroundings, and vice versa. Many species solve this problem by associating different habitats with different stages of life. Both eels, anadromous salmoniform fish and the sea lamprey have different tolerances in salinity in different stages of their lives.

Status

North America: About four in ten North American freshwater fish are endangered, according to a pan-North American study, the main cause being human pollution. The number of fish species and subspecies to become endangered has risen from 40 to 61, since 1989.

Marine Fish Breeding

Very little information has been collected concerning the reproductive habits of marine Angelfish but more information is coming in all the time. One study in 1978 by Moyer and Nakazono describes the reproductive habits of Centropyge interruptus, a species that occurs in harems containing a single dominant male and one to four females. The harems exhibit a pecking-order hierarchical system of dominance. If the male is removed from the harem, the top ranking female will change to the male sex over a period of only two to three weeks. Similar changes have been observed for Genicanthus semifasciatus by Shen and Liu (1975) and for G. lamarck by Suzuki. It appears that harem formation and sex inversion is common among the species of Centropyge and Genicanthus.

Spawning occurs in Centropyge interruptus (Moyer and Nakazono) daily between May and October and seems to be controlled by temperature and light conditions. Spawning will not occur if the temperature drops below 22° C (72° F). The majority of the spawnings take place between 10 minutes before and 5 minutes after sunset unless it was heavily overcast in which case the activities occurred much earlier in the day.

Courtship consists of rapid rushing and circling of individual females by the male. In the final stages before actual spawning the male exhibits a "soaring" display in which he swims high off the bottom above a female and then strikes a motionless pose with all the fines extended. Eventually the female approaches at close range and is "nuzzled" around the abdominal region by the male. This leads to sudden spawning burst in which eggs and sperm are simultaneously released in open water.

The eggs are buoyant and float to the surface where they are planktonic for an unknown time. Moe (1977) reported a hatching time for western Atlantic angelfish eggs of 18 to 30 hours.

Monogamous pairing is the reproductive unit of the large Pomacanthus and many of the Holacanthus angels.

The Pomacanthus demonstrate several types of mating systems, dependent on the density of the population at the particular location. They will form either permanent pair bonds, harems consists of a male defending several females in a small area, or (in at least one species) a large group coming together to spawn. In all systems however, these fish spawn in pairs.

Pairs congregate at the edge of the reef at sunset, usually well above the substrate. They often engage in a courtship display where the male and female swim in a brisk head to tail circling motion. Each pair will spawn and ascend into the water column, swimming together in an arc up to about 7 - 10 feet (2 - 3 metres) above the substrate to expel pelagic eggs at the summit. In the large group spawning system they are quite promiscuous, and males will often try to disrupt the spawning of conspecifics.

Holacanthus (including the Queen Angelfish, the Blue Angelfish, and the Rock Beauty Angelfish):

Adult Queen and Blue angelfish are generally found in pairs year round, so it is assumed that the male and the female have a monogamous relationship. Pairs will spawn by slowly rising up in the water column while bringing their bellies close together, and releasing large amounts of eggs and sperm. A female can release anywhere from

25 to 75 thousand eggs each evening. This can total as many as *ten million* eggs for the duration of the spawning cycle. The eggs are transparent and pelagic, floating in the water column. The eggs will hatch in 15 to 20 hours. At this point the pre-larval angelfish is attached to a large yolk sac, and has no functional fins. The yolk sac is absorbed and during this time the fish develops into true larvae and begins to feed on plankton in the water column. Growth is rapid and 3 to 4 weeks after hatching the fish will reach about 15-20mm and will settle on the bottom.

Butterflyfish - (Chaetodontidae)

Individual butterflyfish species studied up to this time indicate that these fish are gonochoristic, meaning that each fish is either a male or a female,... and they do not change sex!

Marine butterflyfish are not readily sexed as males and females as they do not show different colour displays. There are however, a few species in which the male may be larger than the female. The primary visual difference can be seen in the swollen appearance of a female a couple of days prior to spawning.

Very little information has been collected concerning the reproductive habits of saltwater butterflyfish. They are known to have a variety of social structures that include monogamous pair bonds, harems, and schooling (most likely with group spawning). Reproductive behaviour is variable and is related to their life styles, which in turn is inter-related to the distribution and quantity of food resources.

In the wild, diets vary greatly among the different genera and includes live coral polyps, sessile invertebrates, and zooplankton. Butterflyfish that are planktivores have a food resource that is temporary and unpredictable, lending itself to schooling fish and they are believed to spawn in groups. While butterflyfish that are corallivorous have a stable and predictable food source, lending itself to the defending of territories in a monogamous and/ or a harem structure with pair spawning. Monogamous pairing can be life-long with some genera.

Butterflyfish have a very definite breeding season, it is related to a seasonal change (depending upon the global location) where the temperature of the water either rises or falls to between 73° - 79° F (22.5° - 25.5° C). Other than group spawners, most species spawn at their feeding territory. But there are also some species that do a migratory spawn. These leave their feeding territory at the end of the day and swim out to the edge of the reef where they set up a temporary smaller territory. Here they spawn and then return to their feeding territory in the morning. Females will spawn once per evening.

Courtship displays can be quite energetic with the spawning pair circling each other head to tail, then chasing each other about, and also shooing away any intruders. Spawning occurs with the male behind and under the female as the pair rise up in the water column forming an arc. At the pinnacle of the arc, several feet above the substrate, they releasing their pelagic gametes into the water column. Often times intruding males will attempt to join the spawn, rising and releasing gametes simultaneously with the spawning pair.

The eggs can hatch anywhere between 1 - 30 days. Butterflyfish have a unique larvae, known as Tholichthys, which are covered with bony plates and have a spiny head. These larvae will remain in a pelagic stage for periods ranging from a few weeks to several months, and then finally move into a benthic lifestyle as juveniles.

Marine butterflyfish have not reportedly been spawned successfully in captivity. There are, however, reports of some success in rearing wild collected larvae of some corallivorous butterflyfish. It is hoped these captive reared fish will be adapted to accept aquarium foods, and thus broaden the species selections that can be sustained in captivity.

Clownfish

Many clownfish have been successfully bred in captivity. They are among the first marine species to be bred commercially. Most damsels and clownfish are substrate spawners, which means they lay their eggs on the substrate.

Usually, the male takes care of and maintains the eggs after the fish have spawned. He will keep them fanned and protect them against predators.

To breed them get a pair, or purchase six or more of the type you wish to breed. This will help assure that you get a male and a female of the species. The best success is achieved if they are provided a healthy host anemone that they accept.

After they are mature, about a year or so unless you buy a pair, they will clean off a place near the anemone and lay eggs there. The parents will take care of the eggs by defending them against other fish and will frequently fan them. After about 8 days, the eggs will hatch an hour after dark. This is so they have better chance of avoiding predators, including their own parents! From here the fry are on their own and will need in fusoria or baby brine shrimp several times a day to survive.

It is a good idea to remove the eggs before they hatch and place them in a separate aquarium devoted to raising the fry. Even so, many of the small fry will die before reaching maturity.

There is also an excellent book available on the propogation of clownfish, *Clownfishes*, by Joyce D. Wilkerson. It includes sections not only for selecting your breeders, but also discusses the spawning and rearing of the fry.

Damselfish

Successful institutional breeding of some species of damselfish has been accomplished in India, also some species have reportedly spawned in home aquariums.

Breeding in the Wild

Unlike most saltwater fish species that are pelagic spawners, damselfish just as the clownfish, are substrate spawners which means they lay their adhesive eggs on the substrate. A male damselfish will establish a territory and then clean and prepare an area such as a piece of rubble that has a smoothed-walled crevice, a coral surface, or a rocky ledge for the female to deposit her eggs on.

Once the spawning site has been prepared, the male will then try to attract a female into his territory to deposit her eggs. The male engages in a number of tactics to entice the female to approach involving colour changes and excited swimming movements. Once the female enters the spawning site, she will deposit up to 20,000 tiny oval adhesive eggs on the prepared smooth surface and then leave. The male will then quickly fertilize the eggs.

Each courtship and mating process takes about 10 to 20 minutes and a male may continue to mate with several additional females. In some species the male will tend the batches of eggs by fanning water across them with his fins, often picking out and eating dead eggs, presumably to prevent a fungus from developing that could threaten the whole batch. Some males provide no direct parental care except to protect the eggs from predators. In all cases, the male will aggressively and fearlessly guard his territory and the eggs from intruders, even fish much larger than himself.

The eggs will hatch in 3 to 7 days. The larvae then drifts away as plankton, feeding on the other zooplankton and phytoplankton for several to many weeks, depending on the species. A damselfish can take from 2 to 3 years or even longer to mature, with each species having its own timeframe.

Breeding in Captivity: Reports of the propogation of damselfish has been rather sporadic, though there have been some exciting successes. In India five species have been successfully developed. The Central Marine Fisheries Research Institute in Vizhinjam, Thiruvananthapuram has cultivated the Filamentous- tail Black Damsel *Neopomacentrus cyanomos*, the Yellow-tail Damsel *Neopomacentrus nemurus*, the Blue Damsel *Pomacentrus caeruleus*, the Peacock Damsel *Pomacentrus pavo* and the Indian Dascyllus *Dascyllus carneus*.

Hobbyists too have reported some successes in spawning various damsels in the aquarium. Some of these include the Yellow-belly Damsel *Amblyglyphidodon leucogaster*, Yellow-tailed Damsel *Chrysiptera parasema*, the Blue Devil Damselfish *Chrysiptera cyanea*, the Fiji Blue Devil Damselfish *Chrysiptera taupou*, the Domino Damsel *Dascyllus trimaculatus*, and apparently the Spiny Damselfish, *Acanthochromis polyacanthus*.

The successful propogation of most marine fish are a challenge to even the most advanced hobbyist, and damselfish are no exception. Probably the biggest reason there has not been a lot more reports about the propogation of damselfish in captivity is due to the commitment required in both time and equipment, which of course ultimately translates into cost. This requires a serious commitment, not only to spawn these fish but to rear the fry as well. As damselfish are inexpensive and readily available, there hasn't been a large demand for captive reared specimens.

With an increasing interest in the marine hobby today, many people would like to try their hand at raising these fish. Tto help get started propogating damsels in the aquarium, an excellent book is available on the propogation of clownfish (one of the damselfish genera). It is *Clownfishes*, by Joyce D. Wilkerson and includes sections not only for selecting your breeders, but also discusses the spawning and rearing of the fry.

Hawkfish

The Longnose Hawkfish is also the only one we know of to have been bred in captivity.

Hawkfish in general, are not sexually dichromatic (different sexes cannot be determined by looks). The Longnose Hawkfish, Oxycirrhites typus, is one exception since the male has an additional black edge on the ventral and tail fins.

Hawkfishes generally live in harems with the largest fish of the group being the male. They are what is called protogynous hermaphrodites, which means that if the male is killed or missing, one of the females will change to a male to fill the gap. This is also believed to be the case with wrasses and angelfish.

The hawkfish spawn at twilight. The male sets up a territory which includes 2 to 7 females. Each night around twilight, he visits each female on her coral block looking for the right ones to court. When he finds a suitable female they swim upwards together in an arch releasing the eggs and sperm together.

Not much is known about the eggs at this point, which are now pelagic, but it is supposed that they spend about 3 weeks in this stage. The fact that they are pelagic for quite some time is used to explain why they are so widely distributed about the ocean.

Although they have not been bred in the aquarium (with the exception of the Longnose Hawkfish), they will set up the territories with the male and several females occupying several different coral branches or blocks.

Lionfish

The Shortfin Lionfish is the most often reported as having bred in captivity.

They are haremic, which means one male will spawn with a group of females. After courting for an entire day, the fish will spawn around midnight. They will rise to the top of the water column during and release gelatinous balls of eggs, each ball containing about 2000 eggs.

The egg balls will dissolve after about 24 hours releasing the individual eggs. The eggs themselves will hatch after about 36 hours. The fry will immediately feed on the available plankton. As with most fish it is difficult to keep enough plankton in the water to keep the fry alive.

The above information was obtained from Seascope, Volume 15, Summer 1998, in an article by Matthew L. Wittenrich entitled Lionfish: An Overview

Moray Eels

Everything we have found on the breeding moray eels indicates that they are not commonly bred in captivity for several reasons...

They will not breed until they are very large and have outgrown most aquariums. Also, morays, with the exception of ribbon eels, are what is called 'gonochoristic'. The female will spawn either as a pair or with several smaller males. At dusk the eggs are released and are basically pelagic. The larval stage of the eel lasts for 6-10 months and so makes it difficult to reproduce in the aquarium.

There are several species of freshwater eels, of the family Anguillidae, that spawn far out in the ocean and release up to 20 million eggs per female. After spawning though, the adults die. The young, called "leptocephali" then travel back to the continental shores where they metamorphose into small eels called 'elvers' that then move into freshwater habitats.

Surgeonfish (Tangs)

Some information has been collected concerning the reproductive habits of marine surgeonfish...

Surgeonfish, along with rabbitfish they are gonochoristic (each fish is either male or female) rather than being hermaphrodites. There are few sexual differences, though in some *Naso* species the males are larger than females. For the others sexual dichromatism exists only during spawning.

Spawnings are tied to the lunar cycle, some surgeonfish will spawn on a new moon and others around the full moon. Many surgeonfish are group spawners, with fish coming together from around the reef late in the afternoon, forming a large aggregate. Others such as the *Zebrasoma* species, spawn in pairs.

The courtship consists of males searching out gravid females, and then changing colours and performing a shimmering movement to entice the female to spawn. The pair rises together toward the surface in an arc shaped path, simultaneously releasing their gametes into the open water at the apex of the arc. Males may spawn with several females in a single session, while sexually mature females spawn only about once a month. After hatching, the pelagic larvae subsist on their egg yolk for a couple days and on day four start to feed on plankton. They then begin to develop into a specialised larva, becoming compressed and growing thorns on the dorsal and ventral fins. Their bodies are scaleless and transparent with a silver cast to the abdomen. This post larvae stage is called 'acronurus larva', and is distinct to the Acanthuridae. As they grow the body becomes oval, the spines on the caudal peduncle develop, and the thorns on the fins gradually disappear (except on some of the *Naso* species and on the Blue Tang *Paracanthurus hepatus*).

The planktonic stage will last about 10 weeks after which the young will then settle into a shallow reef. Though the behaviour of the young will vary between species and with the availability of food, many are initially quite territorial. As they mature most species become less aggressive and begin to roam wide areas of the reef in large schools.

Some species of surgeonfish have spawned in public aquariums and there have been a few scattered reports of spawnings in aquariums, but regular spawnings and rearing of the young has not yet been reported.

Triggerfish

Male triggerfish maintain large territories with several females maintaining smaller territories within the male's larger territory.

Each female will dig a pit in the sand within their territory which will serve as a place to deposit the eggs. I don't yet know how long before the eggs hatch, or what the larvae are called.

Genetic Resources of Indian Major Carps, their Distribution and Characterisation

Genetic Resources of Indian Major Carps

The natural genetic resources of Indian major carps come from the network of the Ganga river system, the Sindh and the Brahmaputra river systems in the north and the east-coast and west coast river systems flowing through in the south and central India respectively.

About Indian Major Carps

The major carps of India fall under three genera, *Catla, Labeo* and *Cirrhinus*. Under the genus *Catla*, the species *C. catla*, under the genus *Labeo* fall the species *L. rohita, L. calbasu, L. fimbriatus, L. bata, L. gonius*, and under the genus *Cirrhinus* fall the species *Cirrhinus mrigala, C. reba, C. cirrhosa.* Interspecific hybridisation work has been carried out between the species of the genus *Labeo* and intergeneric hybridisation between the species of the above mention three genera.

Due to their fast growing nature and taste, Indian major carps enjoy a prime position in the Indian aquaculture scenario. These highly prized fishes though originally inhabitants of the Ganga river network in North India and the rivers of Pakistan, Bangladesh, Nepal and Burma, are also transplanted into other rivers in central as well as peninsular India. Besides this, these carps have been imported by many other countries such as Thailand, Vietnam, Laos (Welcomme, 1988).

Catla feeds mainly on Zooplankton at the upper zone of the water body and rohu mostly feeds on periphytonic (phytoplankton) forms found attached to submerged vegetation and other objects occupying the water column, while mrigal and kalbasu feed at the bottom on bottom biota such as tubifex and other blood worms. Thus these carps have mutually compatible and complimentary food habits.

Taking advantage of the mutually compatible and complimentary food habits of these carps, the present day intensive and extensive polyspecies or multispecies culture has been developed from 1970s, which is popularly known as the composite carp culture. This technology has developed and has demonstrated a gradual increase in the fish production from the culture systems from 3–4 t/ha/year to 10–15 t/ha/year.

These technologies are mainly based on manipulation of species ratios to exploit the available natural food in all the niches of the culture medium and supplementary feeding and manuring. Other management measures such as water replenishment, health monitoring etc. are also part of the package.

Carp culture practice which was hitherto confined mostly to the eastern region of the country has become popular almost all over the country with the development of modern aquaculture technologies. Consequently the demand for seed of major carps gradually started increasing. With the increase in the demand, the hypophysation technique has spread all over the subcontinent and became popular among many fish farmers. As a result the production of seed has increased. With the epoch making success in inducing these carps to breed through hypophysation during late 1950s (Chaudhuri and Alikunih, 1957), the dependence on riverine sources for the seed of these carps has been gradually reduced. Now, hypophysation, has been adopted by many of the rural fish farmers to produce the seed in adequate quantities.

However, in the absence of proper breeding plans, this has led to a gradual decline in the genetic quality of the seed. Consequently the negative effect of inbreeding started appearing gradually with the characteristic poor survival and slow growth, besides disease susceptibility of the hatchery produced seed (Ibrahim *et al.*, 1982). All this suggests that it is an appropriate time to act seriously about the genetic improvement of these carps.

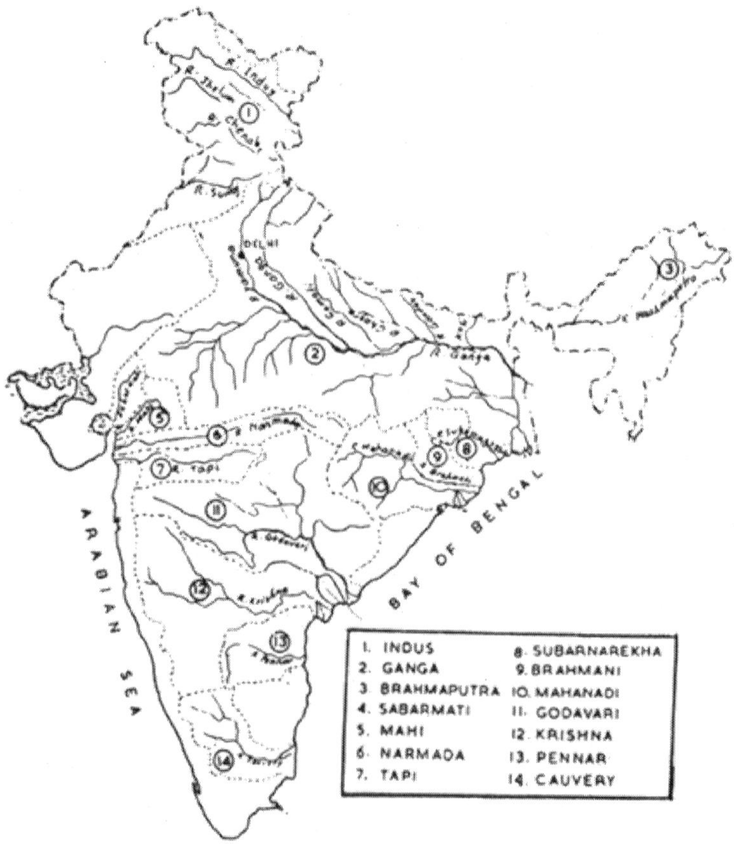

Figure: *Major rivers of India (Jhingran, 1984)*

Spawning Behaviour of Indian Major Carps

All Indian major carps naturally breed in rivers. They also breed in artificially created water bodies which are popularly known as "bundhs". However, Indian major carps though attain maturity in confined or stagnant (pond) waters, do not breed there and need inducement for spawning.

Natural Spawning

Natural spawning of major carps usually coincides with the South-west monsoon in North-Eastern India and Bangladesh and lasts from May to August. In North India and Pakistan, from June to September. In the southern parts of India the spawning season appears to be variable (Jhingran, 1968; Khan and Jhingran, 1975; and Jhingran and Khan, 1979). According to Natarajan and Jhingran (1963), catla breeds only once in a year, as do other Indian major carps under natural conditions. However, recently catla and other species of Indian major carps have been bred more than four times in a season/year under controlled conditions (Gupta *et al.*, 1995).

Regarding the requirements for spawning and the nature of spawning grounds, the observations of various workers differ slightly. Some are of the view that shallow areas which are inundated by heavy monsoon floods form the breeding grounds of these fish (Hora 1945, Hussain 1945 and Khanna 1958). Others viewed that it is only the availability of shallow spawning grounds that is the important factor for spawning (Khan, 1947; Ganapati and Alikunhi, 1950).

Again regarding the nature of the spawning ground, observations differ. In rivers Godavari, Krishna and Cauvery in South India, sections having large submerged rocks or emergent vegetation form the spawning grounds for catla (Chacko, 1946; Chacko and Kuriyan, 1948a; and Menon *et al.*, 1959). Job and Chacko (1948) described as many as 22 spawning grounds. Spawning of major carps was reported over hard and sandy beds and also on rocky submerged embankments of rivers, reservoirs and "bundhs" in Madhya Pradesh.

Riverine fish seed is collected with the help of specially designed, funnel shaped nets (Jhingran, 1968). It consists of two parts, i.e. the net proper and a detachable tail piece locally called as "gamcha" which is fixed to the cod end. The rear of the net tapers to the narrow end. The net is fixed with the broader (mouth) end facing the water current. The spawn/fry are collected at the tail piece.

"Bundh" Breeding

Bundhs are special type of impoundments which are mostly found in the states of West Bengal, Madhya Pradesh, Rajasthan and Bihar. These are of two types, Perennial (wet bundh) and seasonal (dry bundhs). A typical wet-bund is usually located in the slope of a vast catchment area of undulating terrain, having proper embankments and with an inlet towards the upland and an outlet on the opposite side. It has a deeper and vast shallow area which dries up in summer. The deeper portion of the bundh retains water during summer where an adequate stock of carps is maintained. During monsoon when there is a heavy shower, the rain water from the catchment area rushes through the streamlets and submerges a major portion of the bundh while the excess water flows out through the outlet.

On the other hand a "*dry-bundh*" is like a seasonal pond and much smaller and shallower than the *wet-bundh*. For most part of the season the *dry-bundh* remains dry. When sufficient water accumulates with the first showers of monsoon rains brood of mature carp are released into it for acclimatisation. Later when the bundh gets flooded with the next heavy showers, the fishes start spawning and the eggs are collected by means of mosquito net cloth and incubated in hatching *hapas*, which are made up of thin cloth.

Sometimes a few of the female and male brood fish are injected with carp pituitary hormones at a nominal dose of about 5–10 mg/kg body weight of the fish and released along with uninjected brood fish. Both fish that received the hormones and those that have not will also spawn.

Controlled Breeding

In early efforts to induce spawning in the Indian major carps, crude extracts of pituitary gland preferably of the same species (homoplastic glands) were used.

The pituitary hormones are injected either intramuscularly or interperitonially. Depending on the condition of brood fish and the prevailing weather, the total dose of pituitary extract usually ranges from 5 to 22 mg per kilogram body weight in the case of female fish, which is administered in two installments with a six-hour interval. The quantity of first dose is nearly one third of the total dose. The males are given only one dose ranging from 2–3 mg of pituitary per kilogram body weight, at the time of injecting the second dose to the female fish. The dose of pituitary hormones however, appears to be species specific. It also depends on the maturity status particularly in the case of the female brood fish.

In the recent past, due to shortage of quality pituitary glands and the problems in the maintenance of donor fish, many substitutes have come into use to replace pituitary hormones, such as Human Chorionic Gonadotropin (HCG), Salmon gonadotropin (partially purified), Luteinizing Hormone-Releasing Hormone (LHRH) and its Analogues, Gonadotropin Releasing Hormone (Gn RH) etc. However, presently a formula of S-GnRH and domperidone which is marketed in the name of OVAPRIM, is much in use to induce the major carps to breed. OVAPRIM is used mostly for female brood fish @ 0.3–0.4 ml/kg body weight of the fish in a single dose.

With pituitary extracts the female fish will be ovulating within 5–6 hours after receiving the second dose, whereas OVAPRIM takes 6–7 hours to induce ovulation. The release of milt by the males usually synchronized with the ovulation of females. For genetic experiments, stripping procedure is generally followed.

Distribution

The Indian major carps which are also known as the gangetic carps are the natural inhabitants of the Ganga river network, namely the Ganga, the Gomati, the Yamuna and the Brahmaputra and the Indus river system in North India.

The Ganga river system running to a total length of about 8,047 km, besides the major carps, harbours the richest freshwater fauna of India, ranging from mahaseers and the torrential fishes of hills to a wide array of other fishes of great commercial value.

The Brahmaputra river system having a length of 4023 km consists of rich fish fauna of torrential streams in its upper reaches, however the fish are not of much economic value. Its middle and lower reaches however, have several species of carps, catfishes and other commercially important species of fish.

Regarding the Indus system, only part of a small segment is in India, the rest being in Pakistan. The important rivers of this system are the Beas and the Sutlej. These rivers harbour not only the indigenous carps in the lower sections but also the exotic rainbow and brown trouts in their upper reaches.

The main Rivers of the East coast river system in the peninsular India are the Mahanadi, the Godavari, the Krishna and the Cauvery. The Mahanadi naturally has all the species of Indian major carps, which are common in the Ganga system. Rivers Godavari, Krishna and Cauvery have their own several indigenous carp species, which are referred as minor carps such as *Labeo bata, L. gonius, L. fimbriatus, Cirrhinus reba* etc. However, the major carps have been repeatedly transplanted into these rivers and have become established in them, contributing significantly to the fish fauna of these rivers.

The major rivers of the West coast river system are the Narmada and Tapti. Their fish fauna consists of carps, catfishes, mahaseers, murrels, purches, prawns etc.

Distribution of major carp resources also includes that part of the Indus river system which falls into Pakistan after its formation and tributaries of Brahmaputra flowing through Bangladesh.

Distribution of Catla catla (Catla)

Jhingran (1968) described the distribution of *Catla catla*, which starts from the Ganga river network in the north to the Krishna river down south of India, Pakistan, Bangladesh and Burma. It is also found in Nepal (De witt, 1960 as referred by Jhingran, 1968).

The fingerlings of catla, introduced into the Cuddapah-Kurnool canal from river Godavari in 1909 found their way into river Penna and the connected waters in the Nellore district of the state of Andhra Pradesh. Catla fingerlings were also introduced in Cavery river during 1920s and subsequently into the Bhavani (Hornel, 1924). Later the species was introduced into Periyar lake, Powai lake (Chacko, 1948 and Goldschmilt, 1953). The species has become widely established.

Jhingran (1968) has also mentioned that catla fingerlings have been also exported to Israel in 1954 and to Japan, and Mauritius in the 1960s. According to FAO Database on Introduction Aquatic Species (DIAS) catla has been distributed to other countries also such as Zimbabwe, Israel, Bhutan, Philippines, Former USSR, Japan, Sri Lanka, Laos, Pakistan, Malaysia, Thailand, Vietnam and Mauritius.

Distribution of Labeo Rohita (Rohu)

The main source of literature about the distribution of rohu is Day (1878 and 1889). He reported that the species is distributed in freshwaters of Sindh and Punjab (Pakistan), through India, Assam,

Bangladesh and Burma. He also stated that rohu is not found in Madras (South India).

Later studies of other workers (Setna and Kulkarni, 1949) and Alikunhi and Chaudhuri, 1951 and Mishra, 1959) have mentioned the occurrence of this species in many other places, viz., Sabarmati drainage, in the rivers Narmada, Tapti, Godavari, Mahanadi etc. though it is more common in plains of North India.

Its distribution has been also mentioned in other neighbouring countries such as Bangladesh, Burma and Nepal. Khan and Jhingran (1975) have given other details on the distribution of rohu through transplantation. However, according to FAO DIAS besides the above mentioned countries the distribution of rohu has also extended to the Philippines, Former USSR, Japan, Sri Lanka, Laos, Pakistan, Malaysia, Thailand, Vietnam, Madagascar, China and Mauritius.

Distribution of Cirrhinus Mrigala (Mrigal)

The earliest work on distribution of mrigal appears to be Day (1878 & 1889). According to him mrigal inhabits rivers and tanks (water bodies much larger than ponds) in Bengal (undivided which includes Bangladesh) Deccan, North-West provinces, Punjab, Sindh (Pakistan), Cutch (partly in Pakistan and partly in Gujarat, Rajasthan provinces of India) and Burma. Later workers (Alikunhi, 1948 and 1957, and Mishra, 1959) recorded its distribution in the major river systems of India, including river Godavari in the south.

Distribution of Labeo Calbasu

Like the other three major carp species. *L. calbasu* also enjoys wider distribution in almost all those countries namely, Pakistan, India, Bangladesh, Burma, Thailand and also in Yamuna (South China). Day, 1878 mentioned that *L. calbasu* is distributed in Punjab, Sindh, Cutch Deccan, South India and Malabar, from Krishna through Orissa, Bengal and Burma. In India, according to Jhingran (1984) the species is found in streams of Punjab, West Bengal, Orissa and parts of South India. It is more abundant in rivers, above tidal reaches and lakes.

It is an important food fish and at several places it is referred as "lake rohu". It is an important game fish too in tanks. It has been reported that *L. calbasu* thrives better in tanks and lakes than in running waters.

Major Carps in Reservoirs

Reservoir here refers to the impoundments formed as a result of erecting dam on a particular river. There are a number of such reservoirs in India.

Srivastava *et al.* (1985), as referred by Sugunan (1995) have compiled a list of 975 large and medium reservoirs in India, with an estimated area of 1.7 million ha. These reservoirs have major carps as part of their ichthyofauna. All the species of major carps do not seem to thrive equally well in all the reservoirs. Again, there may be fluctuations in the abundance of the same species in the same reservoir.

In Stanley reservoir, in the state of Tamil Nadu, catla was dominating in the fish landings during 1960–'65, but it was reduced to 10% during 1969–'70 to 1972–73. Mrigal, introduced into the Stanley reservoir during 1950–'51, first appeared in the catches during 1957–'58 and contributed over 13% of the catch in 1966–67. However, later catch declined to insignificance. Rohu also showed some initial increase but later almost followed mrigal (Sugunan, 1995). The trend in most of the reservoirs more or less appears the same. The reasons for the dominance of a particular species at one time and the decline of the same species later seem to be due to recruitment failure from the water level fluctuations and predator pressure (Sreenivasan, 1966). Moreover, other aspects like the availability and supply of the required natural food of a particular species may play a role in the establishment and thriving of the species in reservoirs.

Genetic Characterisation of Major Carps

Application of genetics in fish farming is relatively of recent origin. Until recently no attempts have been made to genetically characterise any of the major carp species. No reports are available on whether there exist genetically different populations among the members of a given species of major carps in any of the major river systems mentioned earlier. The main reasons for this seem to be prioritisation of problem/tasks in the fishery sector. As mentioned earlier, the priorities in fisheries, not only in India but also in other developing countries in Asia were different, ever since systematic research has been initiated. The need was to identify the economically important species of fish and study their biology and methods of propagation under controlled conditions. Once this was established, the fishery workers were busy in developing, technologies for carp/

fish farming systems to boost the production. Except for the work on hybridisation, no importance was given to any aspects of genetics what so ever, during these years.

The composite or multispecies culture technologies so far developed are based on species manipulation and application of certain management practices. These technologies, no doubt have boosted the fishculture in India several folds. However, at present it is felt that any further improvement in fish production may not be possible with these technologies and the researcher gradually started realising the importance of other aspects such as genetic quality and improvement of the candidate species by fully exploiting their hitherto untapped genetic potentials. The urgency of acquiring knowledge about the existence of different fish stocks/populations among the different species of the economically most important fish (carps) is now felt. Lack of appropriate and accurate methodologies for genetic characterisation and identification at population species level has also been one of the main constraints to full use of genetic resources. The various methods available earlier, much before the advent of biochemical and molecular (DNA) techniques for stock identification or to study the existence of different populations in a given species were only the morphometric measurements and meristic counts. These methods however, do not provide the degree of polymorphism distinguishable by modern methods especially within species.

According to Jhingran (1968) no distinct races or varieties of catla are known. This author has also given an account of the morphological traits of catla as described by different authors. However, Sinha and Khan (1989) mentioned that in Rihand Dam (Madhya Pradesh), India, the occurrence of three intra-specific populations within catla, each of which can be distinguishable morphologically by short, medium and long pectorals and also by specific ecological roles.

Khan and Jhingran (1975) mentioned that Khan (1972) conducted some studies on rohu collected from two different environments, moats, representing lentic environments and from the rivers Ganga and Yamuna representing lotic environments. The analysis of the data has shown significant differences in each of the characters suggesting that the fishes of moats and rivers belonged to independent stocks. Except for these sporadic reports which are mainly based on morphological and other phenotypic characters, no other reports are available regarding the existence of different populations or races among the species of Indian major carps. Later developments include the cytogenetic, biochemical and molecular genetic methods. Among these, the biochemical genetic techniques, especially the use of isozyme electrophoretic markers, were in extensive use in fishes, through the 1960s and 1970s. However, since isozyme pattern cannot detect point mutations and conservative amino acid substitutions, they do not reveal all of the genetic variation actually present.

At the National Bureau of Fish Genetic Resources (NBFGR) at Lucknow (India) a number of species including major carps have been cytogenetically investigated (Ponniah, 1997). Indian major carps have the same diploid number of 50 chromosomes and no intraspecific variation in chromosomal number has been detected. Similarly, no intraspecific Nuclear Organising Region NOR variation have been observed in both riverine and hatchery populations of *L. rohita*.

A relatively recent advancement made in the area of molecular genetics through developing molecular DNA markers has provided more accuracy for the identification of fish stocks in the form of restriction fragment length polymorphism (RFLP) of mitochondrial DNA (mt DNA) and nuclear DNA. Genetic polymorphism in nuclear DNA can also be detected by using random olegonucleotide primer for polymerase chain reaction PCR amplification followed by gel electrophoresis. This is known as random amplified polymorphic DNA (RAPD), which can be applied to genetic stock identification of quantitative traits loci (QTL) (Padhi and Mandal, 1995a). These techniques are very sophisticated and involve high cost infrastructure, which is often not within the reach of many developing countries. This appears to be one of the main reasons why the studies on stock identification in fishes of many developing areas have not been carried out on full scale until recently. However, use of the DNA technology is gaining momentum since 1990s. Many laboratories and institutions have come up in fisheries and are involved in research in this direction.

Application of Biochemical and Molecular Genetic Techniques

Information on some recent investigations with regard to biochemical and molecular genetic aspects on Indian major carps is discussed below.

Biochemical Genetic Studies

Biochemical genetic studies on Indian major carps through electrophoretic analysis of isozymes was initiated around the mid to late 1980s. These were mainly on the enzyme lactate dehydrogenase (LDH) patterns in developing embryos and in eye lens and red cell haemolysates (Padhi and Khuda-Bukhsh, 1989a,b), Ldh-C gene expression in liver tissue (Rao *et al.*, 1989), malate dehydrogenose MDH pattern in red cell haemolysate and similar electrophoretic patterns in muscle proteins (Padhi and Khud-Bukhsh, 1989b). These studies have shown that the three Indian major carps species namely *Catla catla, Labeo rohita* and *Cirrhinus mrigala*, in many of their biochemical characters were found to be very similar.

Earlier to this, Sahoo (1987) worked on the plasma, haemoglobin and transferrin pattern of two Indian carp hybrids; one produced by crossing male *L. rohita* with female *C. catla* and the other between male *L. rohita* with female *Cirrhinus mrigala*. All the plasma protein bands of both the parents *L. rohita* and *Catla catla* were expressed in their hybrid.

Regarding the haemoglobin pattern the hybrid between *L. rohita* and *C. catla* exhibited a broad intensely stained band which was similar to its parents. Similarly the hybrid between *L. rohita* and *Cirrhinus mrigala* also had a single band closely similar to parents.

With regard to transferrin bands, *L. rohita* has two and *C. catla* has one but their hybrid has inherited the extra band also from *L. rohita*. The hybrid of male rohu and female mrigal has shown two transferrin bands as in both the parents. Thus the pattern of plasma, haemoglobin and transferrin bands in the parent species were inherited by their respective hybrids.

Rao *et al.*, (1989) have studied lactate dehydrogenase isozymes and taxonomic significance of Ldh C-gene in fifty two species of Teleostean fishes including Indian major carps. Liver specific Ldh C-gene expression was observed in *C. catla, Cirrhinus mrigala, L. rohita* and *L. calbasu*, also in many other species including Chinese carps, belonging to family Cyprinidae. Thus, members of the order cypiniformes show a dominance of liver-specific C-gene expression.

Padhi and Khuda-Bukhsh (1989a) studied the lactate dehydrogenase isozyme patterns in developing stages of four species of Indian carps. They found that usually the LDH gene expression during development tends to be similar in closely related species while the pattern may be dissimilar in more distantly related forms. It thus helps to understand the degree of evolutionary kinship among related organisms. The LDH banding latterns in the three species of major carps viz. *C. catla, Cirrhinus mrigala* and *L. rohita* was observed to be very similar and all of them exhibited persistent pattern of three bands in all the developmental stages studied This indicates their genetic closeness (Padhi and Khuda-Bukhsh, 1989a).

Padhi and Khud-Bukhsh (1989b) have also studied malate dehydrogenase isozyme patterns during the development of five species of carps namely *C. catla, L. rohita, L. bata, Cirrhinus mrigala* and *Ctenopharyngodon idella*. Though significant changes in MDH pattern were marked during hatching, in the case of *C. catla* in the form of additional bands, no such MDH activity was apparent in the other species. The MDH patterns however, tended to fluctuate at different pre and post hatching stages in these carps, possibly reflecting differential gene expression (Padhi and Khuda-Bukhsh, 1989b). The number of MDH bands expressed during development, ranged between 1–6 in *C. catla*, 3–4 in *L. rohita*, 2–5 in *Cirrhinus mrigala*. While the number of MDH bands in *L. rohita* did not fluctuate much during pre-hatching stages, differential expression of MDH bands was noticed in other species.

To explain the genetic basis of expression, specific antisera and other confirmatory studies are required. Xanthine dehydrogenase isozymes (XDH) and its distribution in different tissues namely skeletal muscle, heart, liver, eye, kidney and brain and its taxonomic significance have been studied in *C. catla, Cirrhinus mrigala, L. rohita* and *L. calbasu* (Padhi and Khuda-Bukhsh, 1990a). In this study single band XDH phenotype was expressed uniformly in all the six tissues studied in *C. catla*. In *Cirrhinus mrigala* and *L. rohita* it was expressed in most of the tissues. However, in the case of *C. calbasu* and also *L. rohita* multiple XDH phenotypes showing four or five bands were also encountered in certain tissues.

In another study, Padhi and Khuda-Bukhsh (1990a) have found through simple gel electrophoretic study of the surface mucus of catla, rohu and mrigal, conspicuous differences which can differentiate the three species. The authors are of the opinion that a critical analysis of components of surface mucus may also be usefully exploited in population/stock identification of carps. The surface mucus proteins were expressed as seven bands in *C. catal*, three bands in *L. rohita* and five electrophoretically analysable components in *Cirrhinus mrigala*.

The LDH isozymes in embryonic and post embryonic stages of *L. rohita, Cirrhinus mrigala* and *C. catla*, using starch gel electrophoresis. These studies have shown species specific and differential enzyme locus (gene) expression pattern in LDH up to 18th hour of study. The isozymes seemed to be completely active, 36th hour after fertilization and showed electrophoretic patterns very similar to those of the adults. It was also mentioned that comparative analysis of isozymes of the three species has also enabled species identification even during the embryonic stages which other wise would be impossible to identify on the basis of morphological characters only (Basu *et al.*, 1992).

Other studies in the area of biochemical genetics pertaining to Indian major carps are gel electrophoretic patterns of transferrin in the four Indian major carp species, *C. catla, Cirrhinus mrigala, L. rohita* and *L. calbasu* along with 30 other freshwater species of fish. While only one transferrin band was observed in *L. calbasu*, in *C. catla, Cirrhinus mrigala* and *L. rohita* expression of both one band and two band phenotypes were noticed (Barat and Khud-Bukhsh, 1992). Sahoo (1993) studied the plasma, haemoglobin and transferrin pattern in two intergeneric carp hybrids. The hybrids between male *L. rohita* x female *C. catla* and male *L. rohita* x female *Cirrhinus mrigala* have shown a haemoglobin pattern more or less similar to their respective parent species. However, the plasma band profile of the hybrids has shown a mixture of bands of the parent species. Again, the pattern of transferrin bands in the hybrids though resemble the parent species in the number and staining intensity, differed in mobility.

Isozymes studies have also been made to distinguish diploids and tetraploids. Sarangi and Mandal (1996) have made some isozyme studies in one of the Indian major carps namely *L. rohita* to distinguish diploid individuals from tetraploids. In general, intensification of isozyme bands has been observed in tetraploids. The polymorphism in (G6 PD-1 and 2) loci in eye lens, G6 PD-2 in kidney and G6 PD-1 in skeletal muscles and an esterase (EST-1) locus in kidney could be used as reliable marker in identifying tetraploids from diploids (Sarangi and Mandal, 1996).

Chaudhuri and Gopala Krishna (1998) studied tissue specificity and degree of polymorphism of five enzymes systems in *L. rohita* from river Yamuna namely Aspartate transaminase AAT (E.C.2.6.2.2), Malate dehydrogenase MDH (E.C.1.1.1.3.7), Lactate dehydrogenase LDH (E.C.1.1.1.2.7), Malic enzyme (ME) and Glucose Phosphate

Isomerase GPI (E.C.5.3.1.9) enzyme in liver, muscles, heart and brain tissues. It was reported that *MDH, ME* and *GPI* had two loci each in all tissues while *LDH* had trace loci with one specific to the liver. *AAT* with two alleles was the only polymorphic locus among the five enzymes studied.

Gel electrophoretic studies have been also carried out in over thirty five species of Indian teleostean fishes, which include major carps, in relation to plasma protein profiles (Khuda-Bukhsh and Sahoo, 1998). The plasma protein profile of *C. catla* revealed 9 bands, including a cluster of 4 linear bands near the origin. The number of bands in the cluster sometimes varied from 4–6. In *Cirrhinus mrigala* the electropherogram of plasma protein revealed 10 bands, including the cluster of bands near origin which generally contained 5 linear bands but sometimes 6–7 bands were also appeared in the cluster. In the case of *Labeo calbasu* and *L. rohita* 10 bands were observed in each case. (Khuda-Bhukhsh and Sahoo, 1998).

Molecular Genetics Studies

Methods for the rapid isolation of mitochondrial DNA (mt DNA) from fish have been developed (Padhi and Mandal, 1993), followed by mt DNA RFLP investigation studies for genetic stock identification (Padhi and Mandal, 1995), DNA fingerprinting (Majumdar *et al.*, 1997) and characterisation of Mbol satellites (Padhi *et al.,* 1998) have been also carried out in major carps. It has been reported that mt DNA technique enables effective yield of good quantity of mt DNA without the use of ultracentrifugation. The authors claim that this method would be useful for preparing large number of mt DNA samples for studies pertaining to genetic stock differentiation, evolutionary analysis of related taxa and identification of maternity of the naturally occurring fish hybrids. The fish species *C. catla, L. rohita, L. bata* and hybrids of *C. catla* X *L. rohita* were used for this study.

The feasibility of actual application of mt DNA RFLP analysis methods have been tested by Padhi and Mandal (1995b) in Indian major carps for genetic stock identification by using *C. catla, L. rohita* and *L. calbasu* from both riverine (river Ganga) and farm sources. Restriction endonucleage analysis of mitochondrial DNA in *L. rohita* from the river Ganga and hatchery maintained (farm) population revealed polymorphism at the Hind III restriction site. The mt DNA genome size in rohu has been reported to be about 17.8 kb. Padhi and

Mandal (1995b) also reported that *C. catla* and *L. calbasu*, collected from both farm and riverine sources, when compared pairwise, did not reveal polymorphism at the Hind III site within each species. The restriction pattern with respect to Hind III site appears to be species specific in catla, rohu and kalbasu. Further study as suggested by the authors would be useful for stock identification and genetic diversity documentation.

Nuclear DNA RFLP methods were employed to detect inadvertent hybridisation in major carp hatcheries when "mixed spawning" of all species Indian major carps is practised (Padhi and Mandal, 1996). It was reported that when southern blots of genomic DNA of the mixed spawned individual offspring digested with EcoR 1 were probed with radiolabelled Xenopus r-DNA the hybrids of catla-mrigal and rohu-mrigal could be detected easily. However, in the case of catla-rohu hybrids caution had to be exercised as the fragment sizes of higher molecular weight diagnostic bands in these species did not vary much. It was also reported that ribosomal DNA fragments were inherited biparentally in catla-rohu, rohu-mrigal and catla-mrigal hybrids. An earlier study on ribosomal DNA RFLP indicated that catla, rohu and mrigal were polymorphic at the EcoR 1 site in ribosomal DNA.

Recently, Majumdar *et al.*, (1997), Basavaraju *et al.*, (1997) and Padhi *et al.*, (1998) have reported on DNA fingerprints in *L. rohita* and *C. catla* by using Bkm 2 C (8) and M 13 multilocus probes. In *L. rohita* and *C. catla*, M 13 gave fewer bands, though the level of polymorphism was higher. This indicates the usefulness of M 13 as an effective DNA fingerprinting probe (Majumdar *et al.*, 1997).

It was also mentioned that although Bkm 2 (8) and M 13 show similar hybridisation patterns, they detect different alleles. M 13 gave only a few bands in Indian major carps which were reported to have clustered in 2–4 Kb size ranges when three different enzymes were used to generate such RFLPs, whereas Bkm 2 (8) detects a large number of bands in the size range of 2–20 Kb.

It was further reported that DNA fingerprinting analysis in *L. rohita* and *C. catla* showed similar patterns with the same probe and enzyme combination. Basavaraju *et al.*, (1997) developed and used four tetranucleotide microsatellite probes to differentiate between broods of catla from the key hatcheries namely, the Tunga Bhadra Dam (TBD), the Bhadra Reservoir Project (BRP) and the Kabini

Reservoir Project (KRP) and from out of state wild and hatchery stocks. This analysis was effective in determining the overall levels of genetic variation between strains in terms of number of alleles and average heterozygosity and in detecting differentiation between strains.

These studies also revealed that despite having a common source, the hatchery populations in Karnataka had differentiated from each other as a result of genetic drift or unconscious selection. It was however, reported that there were significant differences in the number of alleles or in the average heterozygosity between the hatchery and wild caught stocks (Basavraju *et al.,* 1997).

Characterisation of Mbol satellites in *Cirrhinus mrigala* has been reported by Padhi *et al.,* (1998). These workers claim that they have cloned and characterised a highly reiterated, tandamly repeated and A + T rich Mbol DNA fragment in *Cirrhinus mrigala*, with a monomer size of 266bP. It has been reported that in *Cirrhinus mrigala* the Mbol fragment is species-specific and absent in *C. catla* and *L. rohita*. They have also studied the inheritance of the *Cirrhinus mrigala* specific Mbol fragment in two intergeneric hybrids, *Cirrhinus mrigala* X *L. rohita* and *Cirrhinus mrigala* X *C. catla*, when the radiolabelled 32 fragment was hybridised to the Mbol-digested nuclear DNA of *C. catla, Cirrhinus mrigala, L. rohita* and the two hybrids on the southern blot, the *Cirrhinus mrigala* Mbol ladder was found in both the hybrids suggesting that the Mbol satellite in these hybrids was inherited uniparentally (Padhi *et al.*, 1998).

Aneuploidy of a karyotype is due to the loss or gain of one or two chromosones but not due to the loss of an entire chromosone set. Triploids in which all the tree chromosones sets were present were found to be fertile and those which showed chromosome deletion and breakage (aneuploid state) were reported sterile.

Genetic characterisation of the founder stocks of *Labeo rohita* from river Ganga, Gomati, Yamuna, Sutlej and river Brahmaputra and also the farm stock from CIFA, which are being used for selective breeding purpose is being carried out. Results of the preliminary investigations have indicated adequate genetic variability (mean heterozygosity 0.047 to 0.079) similar to wild stocks and no genetic contamination was detected (Kuldeep Lal *et al.*, 1998, unpublished).

Until recent past particularly in India aspects like the geographical distribution and biodiversity of fishes were given priority to estimate the extent of distribution and availability or occurrence of different species. However, after realising the important role of genetics, many a fishery researches too felt the need to exploit the genetic potentials of several commercially important species of fish. To achieve this, information on genetic diversity or divergence, both between and within species is essential. Until recently no attempt has been made with regard to Indian major carps in this direction. However, as discussed in this section, researchers in some of the Asian countries particularly in India have initiated genetic characterisation of these carps by applying both biochemical and molecular genetic techniques. Most of these studies are an attempt based on isozyme pattern and some on mtDNA or DNA fingerprinting by using RAPD or RFLP techniques. These preliminary studies with regard to the genetic status of the hatchery and wild stocks, the expression of C-gene pattern in catla, rohu, mrigal and kalbasu, and the pattern of haemoglobin, plasma and transferrin in major carp hybrids and their respective parent species though provided some information, did not indicate any thing which can focus some light on the genetic diversity or divergence either between or within the major carp species.

Much is yet to be done to find out the genetic divergence of these economically most important Indian carps. As mentioned, some methodologies are now available to proceed further in characterisation and differentiation, to find out whether there exist different populations/stocks/strains among and within the species and the extent of genetic diversity a variation of these carps. Now the research in this direction has been already initiated in India and is in progress.

Present Status of Major Carp Germplasm Resource in Relation to Environmental Changes and Habitat Modifications

It has been reported that there is considerable depletion in the fish populations of almost all economically important species, including major carps in all the rivers (Das & Barat, 1990). This appears to be mainly due to man made stresses by over fishing, more due to greed for profit making business than simply to make a living by subsistence fishing. Besides, there are also other factors like construction of dams and hydro-power plants etc. which prevent the fish from migrating to their natural breeding and feeding grounds. Pollution of aquatic environment mainly due to heavy industries is the concern of the day. The destruction of fish habitat by alterations of river systems, land development, increased water abstraction etc. is very clear (Jhingran,

1984). Mass mortality of fish in seas and rivers is not very uncommon these days. Consequently, several fish species are feared to be at various stages of threat, such as rare, vulnerable, endangered etc. (National Seminar on Endangered fishes of India, Allahabad, (UP), India, 25–26 April, 1992).

According to Ponniah (1997), rohu, catla and mrigal may not be under threat but they are prone to loss of genetic diversity and variability due to extinction of genetically distinct wild populations, escape and ranching of hatchery seed and competition from exotic carps.

Consequent to massive introduction of cultured fish, displacement of wild stocks has been reported and high levels of introgression of the altered hatchery fish genome into wild stock has also taken place. Escape of farmed fish and breeding with wild populations may result in decreased fitness of wild stock.

Another aspect, which may also pose a problem is the growing trend of replenishing the rivers, lakes and reservoirs with hatchery produced fish seed in many states in India (Mishra and Raman, 1993) and even in the neighbouring countries like Bangladesh. This should be given proper thought as most of the hatcheries produce seed without following any genetic resource management (Eknath and Doyle, 1985) and as such it may be disastrous from a genetic view point. Deviations brought about in the gene pool of survivors to that of the original populations effected by pollution will ultimately be reflected in future populations (FAO, 1980). There is likelyhood of genetic deterioration in populations which suffer drastic depletions. Population size is the single most important factor in sustaining a high level of genetic variation within a population (Frankel and Soule, 1981). Therefore, it is important to restore the size of populations whenever trends of decline are indicated. Das and Barat (1990) have dealt with fish conservation methods that are necessary following habitat degradation. Obviously, the various causative factors inducing stresses on fishes need to be eliminated or reduced as a first step. However, it appears that very little or nothing can be done with regard to alterations of river systems and increased water abstraction when the greater interest of the nation is put ahead of the interest of fish or local groups of fishers or farmers (Das and Barat, 1990).

In situ and *ex-situ* conservation methods have been also often suggested for the restoration of the original gene pool of the species in question. The major advantages of *in situ* conservation according to Das and Barat (1990) are i) continued coevolution in the wild which provides the breeder with a dynamic source of resistance to the changes in the environment is not possible in *ex situ* conservation ii) maintenance of a carp relative *in situ* permits the breeder to study its ecology and to obtain data that can assist in the selection of germplasm that might otherwise be over looked.

In view of the deteriorating aquatic environment, the first necessary step taken should be the conservation of the germplasm resources of at least all the economically important species either in *in situ* (in the natural environment). *Ex-situ* conservation (Cryopreservation and gene banking) can support *in situ* measures, but should not replace them (Pullin *et al.*, 1991). This has been also stressed in many seminars/symposia or workshops during the past (Jhingran and Gupta, 1989; Nasar and Haque, 1989) and during the National Seminar on Animal Genetic Resources and their conservation, held at Karnal (Haryana State) India from 22–23 April, 1993.

Conservation of genetic stocks, both wild and cultivated species is important because their genetic diversity has been developed through thousands of years of evolutionary process (Swaminathan, 1984). Selected gene pools or genetic stocks having sufficient genetic diversity and their *in situ* preservation provides the genome the scope of acclimatisation with the changes in the environment.

Ex situ method enables the preservation of genotypes in large quantities in an economical way without much management problems. However, this method may also have its disadvantages. For example, in the long run, genome preserved for several decades, when it is revived, may find the prevailing environment hostile. In the case of animals unless both male and female gametes or embryos could be cryopreserved, full advantage of *ex-situ* conservation can not be derived, without further genetic manipulations (Mc Andrew *et al.* 1993).

Both *in situ* and *ex-situ* conservation methods, no doubt have their advantages as well as disadvantages as discussed above. It may be possible to overcome such situations by following rather a cyclical approach through the alternate use of these two methods by intermittently reviving such *ex-situ* preserved germplasm and maintain them again in *in situ* and in the process to expose and acclimatise the genome to the prevailing environment.

The development of molecular genetics offers another novel and great potential method of preserving genetic materials. The genetic material of individuals having the most desirable and heritable traits can be preserved in the form of genomic libraries.

Hybridisation

Hybridisation in nature among fish species has been reported by several workers (Hubbs 1955, and Slastenenko 1957). Indian major carps are no exception.

Hybridisation of Indian Major Carps in Nature

The occurrence of fish hybrids in nature has been reported from time to time by several fishery workers. Slastenenko (1957) has given a list of natural hybrids of the world which suggests that species belonging to the cyprinid group are more prone to interbreed than other groups of fishes and Indian major carps belong to this group.

Natural occurrence of both interspecific and intergeneric hybrids of Indian major carps has been reported mostly from reservoirs, formed due to the construction of dams across rivers. These hybrids were also reported from dry "bundhs". Occurrence of rohu-catla hybrid has been reported from the perennial irrigation tanks (reservoirs) in the state of Madhya Pradesh in India. A detailed study of the taxonomic characters of this hybrid has been made by Desai and Rao (1970). Occurrence of catla-rohu hybrids has been reported from Adhartal lake also in Madhya Pradesh. On the basis of the morphological characters and position of the pectoral and anal fins, the parental species of these putative hybrids have been identified as female catla and male rohu (Tripathi *et al.*, 1974). Natural hybrids of similar type have been also reported by Natarajan *et al.* (1976) from the Rehand reservoir in the state of Uttar Pradesh. The impoundment with an area of 46,538 ha has a mean depth of 25m. The dominant ichthyofauna in the reservoir is *Catla catla* which contributes as much as 98% of the fisheries.

The hybrids as reported by the authors possess similar traits but differing traits which reflect the broad spectrum of character variations in hybrids. These natural hybrids as observed by Natarajan *et al.* (1976) have a general appearance of *C.catla* with a conspicuously smaller head than catla but with a deeper body than rohu. The shape of the mouth was more similar to rohu with infringed lips. The hybrid reported from Madhya Pradesh (Desai and Rao, 1970) on the contrary had feebly fringed lower lip and the mouth showed one rudimentary maxillary pair of barbels. The morphometric traits (ratios) of these natural hybrids more resembled catla in respect of length of head/ diametre of eye and length of fish/length of body, whereas, in respect of length of head/height of head ratio, the hybrid resembled rohu, the intermediate traits being the ratio of length of fish/depth of body. The hybrid has also showed intermediate traits in respect of meristic counts of dorsal and pectoral fin rays. Natarajan *et al.* (1976) have given the full details of morphometric and meristic characters of these hybrids.

Occurrence of other carp hybrids like *Labeo fimbriatus* (Bl) x *L. gonius* (Ham.) and *L.calbasu x C.catla* in Rangwan reservoir in Uttar Pradesh has been also reported by Prasad (1976). He suggested that these hybrids form good substitutes for *L. gonius* and *L. calbasu* in reservoirs. Similarly, the hybrids of catla-rohu and rohu-catla would also form a good substitute for rohu in reservoirs. Natarajan *et al.* (1976) mentioned that the studies carried out in respect of the ecology and fisheries of Freshwater reservoirs in Bhavanisagar (Tamil Nadu), Nagarjunasagar (Andhra Pradesh), Govindasagar (Himachal Pradesh), Rihand (Uttar Pradesh) under the all India Co-ordinated Research Project and also D.V.C. reservoir (Jhingran and Natarajan, 1969) and Tungabhadra (David *et al.*, 1969) have shown that the performance of *Labeo rohita* in reservoir ecosystems was not much satisfactory when compared to that of *Catla catla, L. calbasu* and *Cirrhinus mrigala*. The catla-rohu hybrid, having wider spectrum than rohu was suggested to be a good substitute to rohu. *Ceratium* sp. constitutes a very important fraction among phytoplankters in many reservoirs. This ecological niche is not utilised properly. Introduction of Catla-rohu hybrid may help boost the fish production from reservoirs through the proper utilisation of the above mentioned niche.

Several causes have been suggested to account for the occurrence of hybrids in nature. Hubbs (1955) holds the view that hybridisation in nature is facilitated when there is scarcity of one species and dominance of an allied species in close proximity. In "bundh" breeding occurrence of natural hybrids may be due to congestion in the spawning ground. Because of the limited space, there is very likelihood of ova of one species getting accidentally fertilized by the sperm of another species. Particularly, in the case of Indian major carps with very compatible genomic structure (Zhang and Reddy, 1991), it is all the more easy and frequent to encounter natural hybrids when bred together in a relatively congregated condition.

Artificial/Experimental Hybridisation in Indian Major Carps

In any genetic improvement programme, it is necessary to be able to breed the candidate species artificially under controlled conditions. Soon after the success of induced breeding of the Indian major carps, work on genetic improvement of these carps was initiated in the late fifties starting with simple interspecific and intergeneric hybridisation to produce and evaluate their useful traits for aquaculture, since hybridisation was regarded to be one of the methods for combining desirable traits of selected species.

Accordingly, hybridisation work was carried out between different species of a genus (interspecific) and species of different genera (intergeneric) among the Indian major carps and Indian and Chinese major carps, including the common carp (*Cyprinus carpio* variety *communis*, Linnaeus,) to study whether the hybrids acquired any useful traits from their respective parent species, such as smaller head, wider body and more flesh content as in rohu-catla hybrid, pond breeding habit as in *Cyprinus carpio*, or advan-tageous feeding habits. Altogether six interspecific and 13 intergeneric hybrids have been produced among the four species of Indian major carps belonging to the three genera, viz. *Catla, Labeo* and *Cirrhinus*. Some of the interspecific and intergeneric hybrids of Indian carps and also between Indian carps and common carp (*Cyprinus carpio var. Communis*) which were observed to possess some useful traits from aquaculture point of view are described here. Information regarding other hybrids involving Indian major carps is given in the Annex.

Interspecific Hybrids

As mentioned above, among the six interspecific hybrids produced and evaluated, characteristics of some useful interspecific hybrids were found.

Characteristics of the Interspecific Hybrids of Indian Major Carps

Almost all the hybrids, either interspecific or intergeneric, produced among Indian major carps in general exhibited intermediate traits to their parental species. The percentage of fertilization was very high reaching about 94% and development of eggs was normal. The offspring of both the hybrid crosses were highly viable and their growth rate was superior to their slow growing parent viz. *L. calbasu* (Chaudhuri, 1971). Hybrids of both the crosses were found to be fertile and attained maturity within two years.

L. Fimbriatusx L. Rohita (Fimbriatus-rohu) Hybrid

Basavaraju *et al.* (1990) made a detailed investigation regarding the growth, food conversion ratio and protein efficiency ratio of the hybrid produced between fimbriatus (female) and rohu male. The growth of the hybrid was evaluated in comparison to the parent species. In an 182 day rearing experiment the hybrid was reported to attain 247.8g as compared to 324.5g in rohu and 161.7g in fimbriatus. The growth in the hybrid was 34.75% higher than that of fimbriatus and 23.55 less than rohu. The hybrid also showed the highest survival, being 90% compared to rohu with 86.67% and fimbriatus, 76.67%. In terms of percentage weight gain, specific growth rate and feed utilisation, the hybrid performed better than both the parental species.

Figure : *Interspecific hybrids cross between female* L. rohita *and male* L. calbasu a. L. rohita, b. *Hybrid,* c. L. calbasu

Figure : *Reciprocal interspecific hybrid-cross between female* L.calbasu *and male* L. rohita a. L. calbasu, b. *Hybrid,* c. L. rohita

5

Estimation of Mortality Rates

This chapter dealt with growth, the positive aspects in the dynamics of a fish stock. This chapter deals with its negative counterpart, the death process or mortality. The growth was described with the aid of a model and a number of parameters, and so is the death process going to be. The key parameters used when describing death are called the "*mortality rates*".

This chapter deals with the definition of mortality, and it introduces some of the methods to estimate mortality rates which require input data from random samples representative of a certain part of the population, namely the exploited part. These methods do not require estimates of total catch from the population. On the other hand, they do not provide estimates of population size. Methods dealing with those aspects will be introduced.

The easiest way to describe the change in numbers in a fish stock is often to follow the fate of fish spawned at approximately the same time, a cohort. We shall consider the mortality of a cohort as composed of the mortality caused by fishing and that due to all other causes lumped together as the "*natural mortality*". The latter covers events such as predation, disease and deaths due to old age.

The Concept of a Cohort and Some Basic Notation

A "*cohort*" is a batch of fish all of approximately the same age and belonging to the same stock. The concepts "day of recruitment of a cohort" and "recruitment" were introduced. In all the following derivations we assume (with Beverton and Holt, 1957) that a cohort consists of "average fish" only. This means that all fish of a cohort are assumed to have the same age at a given time so that they all attain the "*recruitment age*", Tr, at the same time. In this chapter we similarly used the average length of a cohort to describe growth. In the context of mortality rates we are interested in the number of survivors from a cohort as a function of time.

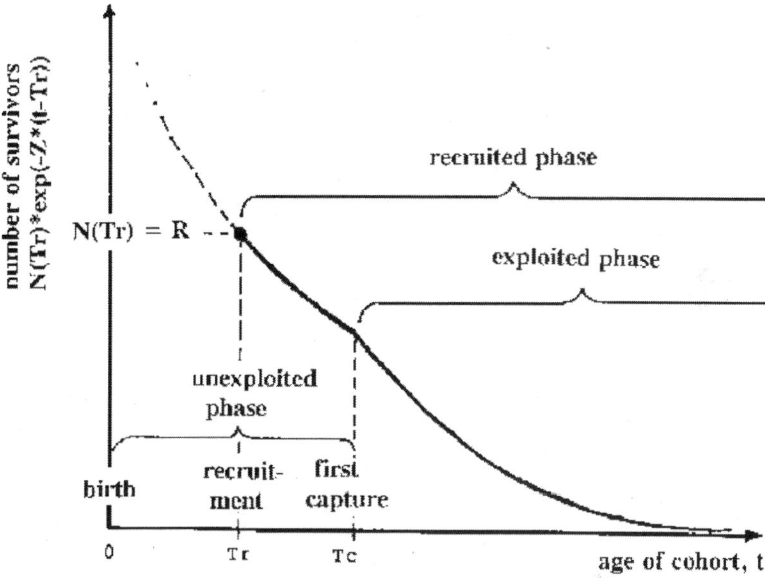

Figure: *Basic features of cohort dynamics*

The symbol "N(t)" is used to designate the "*number of survivors from a cohort attaining age t*". The age is usually measured in units of years. Thus, N(Tr) is the "*number of recruits*" to the fishery. Often the symbol "R" is used to designate the "*recruitment*", R = N(Tr). The actual choice of Tr is not critical, since all calculations are based on relative ages (age differences).

In many applications we do not need to define the recruitment age. Tr is the minimum age at which the fish can enter the fishery, i.e. become liable to encounter with fishing gears (Beverton and Holt, 1957). The age at which they actually enter the fishery, Tc, is dependent on the mesh size. Tc is called the "*age at first capture*" and marks the beginning of the "*exploited phase*".

Figure illustrates the basic features of cohort dynamics. Due to mortality (fishing or natural causes) there is a continuous decrease in the number of survivors. At birth the cohort has age zero. From age 0 to Tr the cohort is in the "*pre-recruitment phase*". In the present context we are not concerned about what happens before age Tr. After age Tr the fish may be caught if a suitable small meshed gear is used.

At age Tc the fish start to be caught with the mesh size actually in use. By definition we must have Tr ≤ Tc.

In some applications we consider several cohorts at a time as illustrated by figure, where a situation with two cohorts per year during a period of two years is depicted. Figure is a parallel to the multi cohort case for growth which was illustrated in figure.

If more than one cohort is considered at a time, the symbol N(t) is not sufficient, and some index to indicate the cohort is then required. However, in the following only one cohort at a time is usually considered, so the short symbol N(t) will do.

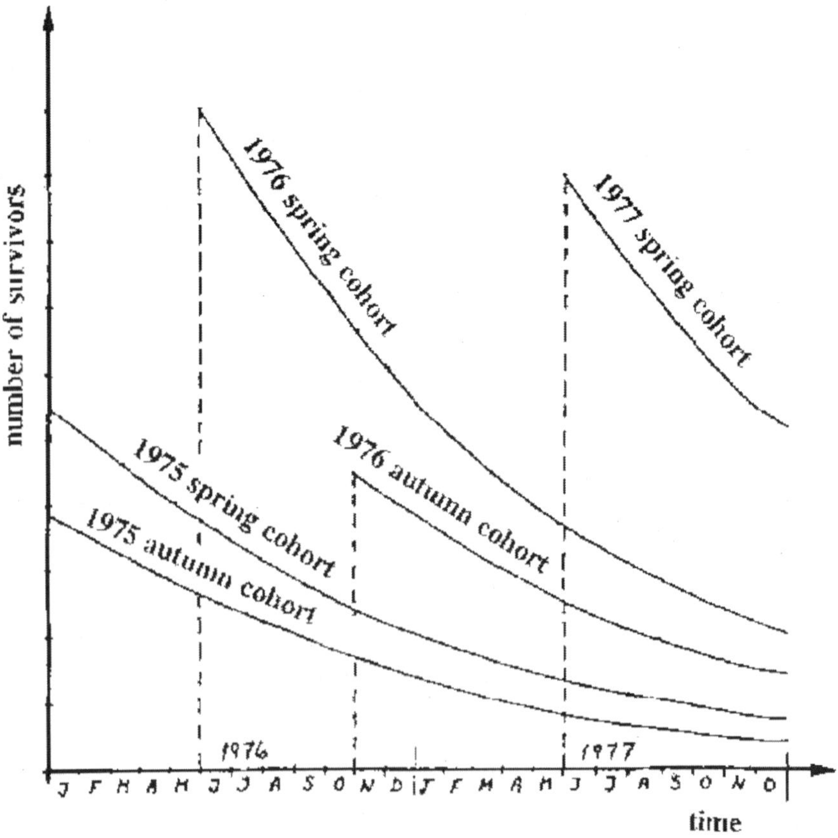

Figure: *Illustration of the situation with more than one cohort*

Estimation of Z from Catch Per Unit of Effort Data and the Concept of the Catchability Coefficient

The total mortality coefficient Z can be estimated when estimates of the number of fish in the cohort are available for two different moments during its exploited phase, t1 and t2. To calculate Z from such data may be rewritten:

$$Z = \frac{1}{t2 - t1} * In\frac{N(t1)}{(t2)}$$

For the estimation of Z with this formula, it is not necessary to know the absolute values of N(t1) and N(t2), only their ratio is required. Thus, if only an estimate of some quantity proportional to N(t) is available above equation can still be applied.

The catch per unit of effort (CPUE), for example, the numbers caught per trawl hour, is a quantity which can be assumed proportional to the number of fish in the sea, N. Intuitively, it is not hard to accept that if there are twice as many fish in the sea, twice as many will be caught per fishing operation. Let CPUE(t) be the number caught of a cohort per unit of effort at time t, then mathematically, the assumption is expressed:

$$CPUE(t) = q*N(t)$$

where q is a parameter called the "*catchability coefficient*". The more efficient the gear is, the higher the value of q because q is a measure of the ability of the gear to catch fish.

CPUE Data from Research Surveys

Assume that we have obtained CPUE data for a certain species with a research vessel during a trawl survey in different periods, t1 and t2, using the same gear without any modifications and thus with a constant catchability coefficient, q. In that case it follows from above equation that:

$$\frac{N(t1)}{N(t2)} = \frac{q * N(t1)}{q * N(t2)} = \frac{CPUE(t1)}{CPUE(t2)}$$

Inserting the above into Eq. 1 gives:

$$Z = \frac{1}{t2 - t1} * In\frac{CPUE(t1)}{CPUE(t2)}$$

Above equation can be used to obtain an estimate of Z from time series of CPUE data from research surveys if age compositions have been obtained, either from reading ring structures in hard parts or from length-frequency analysis. When age data are available, the numbers of fish caught per unit of effort of each age group (cohort) can be determined directly.

So when such numbers are available for a time series the numbers of a certain cohort present at different moments can be used in above euqation. Recall that the results of a length-frequency analysis, e.g. the Bhattacharya method also include estimates of the numbers of fish belonging to each cohort.

If such numbers are available for the same cohort at different times, then the progressive decline in numbers represents the total mortality and equation can be applied to obtain an estimate of Z. An illustration of this method is provided, based on the data presented in table.

CPUE Data from Commercial Fisheries

Sometimes CPUE is derived as the average value over a longer period. This is the case when data are collected from a commercial fishery, where data are most often given on a quarterly or yearly basis.

The model to be used is mathematically equivalent to above equation, but conceptually it is slightly different. The parallel to corresponding to a longer time period is:

$$\overline{CPUE}(t1,t2) = q\overline{N}(t1,t2)$$

This is the model relevant to data on commercial fisheries. The mean CPUE, $\overline{CPUE}(t1,t2)$, is usually calculated as the catch of a cohort during the period from t1 to t2 divided by the effort during that period. $\overline{N}(t1,t2)$ is the average number of survivors during the period from t1 to t2.

The theory behind equations will be used several times in the following chapters. We therefore will discuss the reasoning leading to these expressions.

Let f be the effort during one year (t2-t1 = 1). 4 may then be written:

$$c/f = q^* \overline{N}$$

When t2-t1 = 1 year we have a special version:

$$C = F^* \overline{N}$$

Inserting Eq. 6 into Eq. 5 gives

$$F = q^*f$$

Above equation can be shown to apply to any time period [t2,t1]. It is intuitively easy to accept. It says that the more effort (e.g. boat days) goes into the fishery, the higher will be the fishing mortality, or the fishing mortality is proportional to the effort.

As above equation is so simple we could have taken that as the starting point for the derivation of figure illustrates in a simplified system, consisting of 2 million fish uniformly distributed over a coastal area. The figure shows only 200 fish (1/10000 of the system).

The rectangle labelled "A" indicates the area swept by the trawl in, say, one hour = Δ t. The catch in this short time, Δ C1, consists of six fish. For simplicity, we assume that no fish dies due to natural causes during the towing period and that all fish encountered by the trawl are retained.

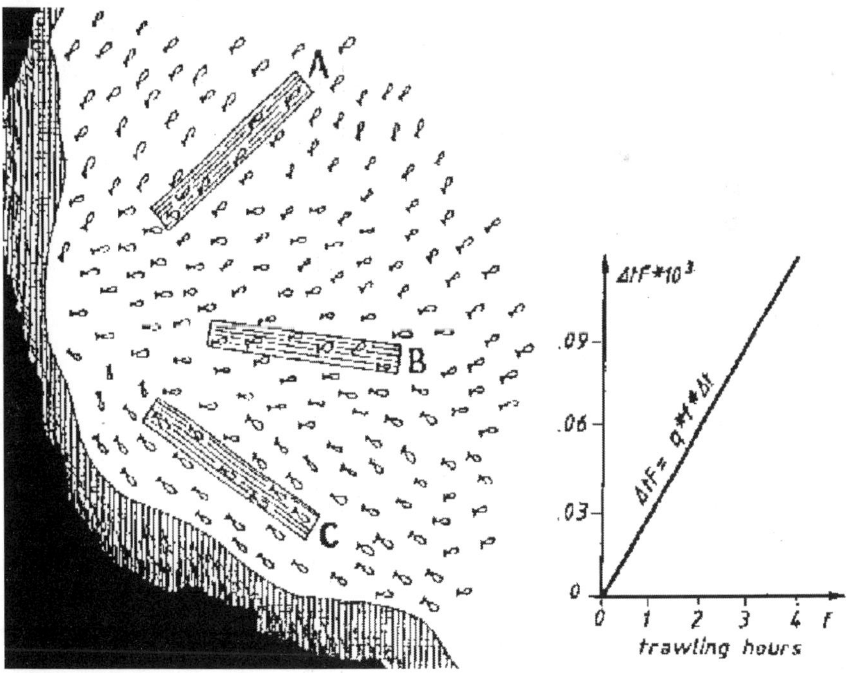

Figure: *Illustration of F = q*f, in the case of a catch, Δ C, which is small compared to the population,* $\overline{N}$

The catch, Δ C1 created by a trawl haul must depend on the duration of the haul, Δ t, and on the abundance of fish in the sea, $\overline{N}$. The simplest model reflecting these intuitive assumptions is that of proportionality:

$$\Delta \text{ C1} = \text{q}^*\Delta \text{ t}^* \overline{N}$$

If we can accept this model for one haul there should be no problem in accepting it also for two hauls (say haul A and haul B):

$$\Delta \text{ C2} = \text{q}^*\Delta \text{ t}^*\text{f}^* \sum_{t} CRR(s,t^*,^*,^*,i) = CRR(s,^*,^*,^*,^*,i)$$

where the effort, f = 2 hauls. Extending the model further with haul C to three hauls still gives no problems, because the abundance, $\frac{Y(t,t+\Delta t)}{R} = F * \frac{N(t)}{R} * W(t) * \Delta t$ will remain approximately constant as the three hauls will only remove a small part of the total population.

Combining Eq. 8' with Eq. 10 gives (when Δ C is small):

$$\Delta \text{ C} = \text{q}^*\Delta \text{ t}^*\text{f}^*$$

$$Y / R = F * \exp\left[-M * (Tc - Tr)\right] * W_{\infty} *$$

$$\left[\frac{1}{Z} - \frac{3S}{Z+K} + \frac{3S^2}{Z+2K} - \frac{S^3}{Z+3K}\right]$$

$$= \Delta \text{ t}^*\text{F}^* \frac{10\,000\,000\,000}{10\,000\,000} = 100 \textit{grams per recruit}$$

which implies that q*f = F. Thus Eq. 4.3.0.7 is intuitively acceptable when Δ C is small.

It is less clear how effort and fishing mortality are related when we consider a longer time period during which the abundance cannot be assumed to remain approximately constant. However, if a number of assumptions are met the equation also applies to cases where the catch, C, is large relative to the abundance,

$$Y / R = F * 147.8 * \left[\frac{1}{F+1.1} - \frac{1.9244}{F+1.47} + \frac{1.2344}{F+1.84} - \frac{0.2639}{F+2.21}\right],$$

and where the time period, Δ t, is long (say one year). In reality we often assume that these assumptions are being met. The above reasoning applies in the case where the fish are uniformly distributed over the fishing grounds. However, this is rarely the case. For schooling fish (pelagics) the argument is simply wrong. When shooting the net at random the catch of a schooling fish species might be zero in nine out of ten tows, yet one tow might yield a very high catch. For demersal fish species the reasoning may be approximately correct although demersal fish also tend to aggregate in certain areas, which may or may not also be the fishable areas. The reasoning illustrated by figure given is behind the "*swept area method*". That the reasoning is based on a relatively small catch and very few effort units, however, does not limit its applicability. Now consider a system during a longer period. Figure shows such a (hypothetical) example, where in total more than half of the original number of fish are removed by fishing. Note that one unit of effort (one trawl haul) produces the same fishing mortality Δ t*F, independent of the number of survivors:

$$Y/R = 0.5 * 147.8 * \left[\frac{1}{0.5+1.1} - \frac{1.9244}{0.5+1.47} + \frac{1.2344}{0.5+1.84} - \frac{0.2639}{0.5+2.21}\right]$$

$$= 0.5 * 147.8 * 0.0785 = 5.8 = 5.8\,grams\ per\ recruit$$

where » means "approximately equal to". One unit of effort (one trawl haul or tow) takes the same fraction from the stock. The fishing mortality is independent of the number of survivors, i.e., all hauls give equal Δ t*F, Note also that Δ t*F will remain constant only if the fish redistribute uniformly over the area after each removal by fishing.

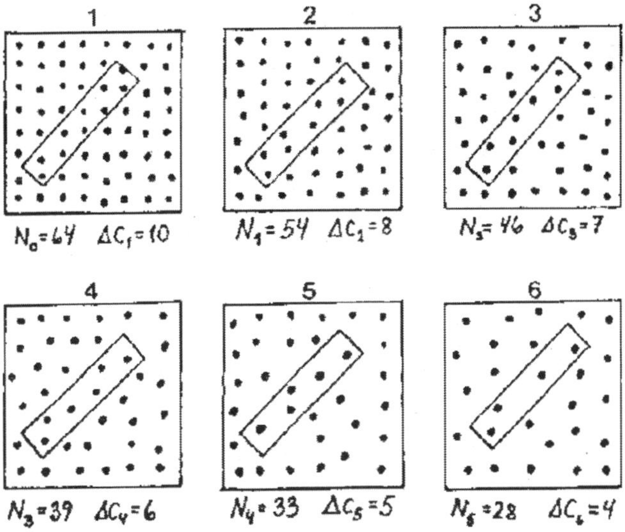

Figure: *Illustration of equation, F = q*f, applied several times to a system during a longer time period. (For further explanation)*

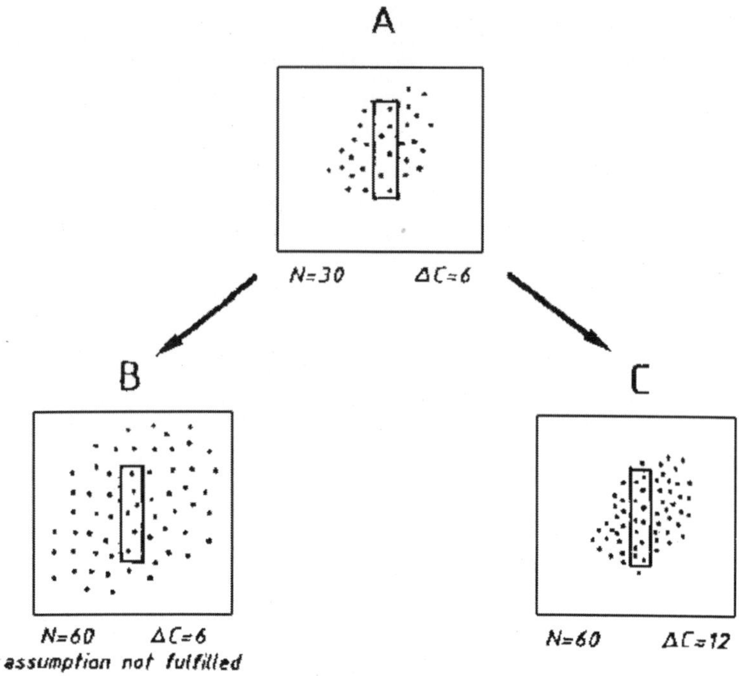

Figure: *Illustration of one of the assumptions behind F = q*f.*

Figure A and B show an example in which F = q*f, is not applicable, *viz.* when an increase in numbers produces also an increase of the area occupied by the stock.

Then one effort unit (one trawl haul) would remove a smaller fraction of the stock and thus produce a smaller fishing mortality. Pigs. A and C illustrate the way a stock must behave to conform, namely that the area occupied remains constant for different stock numbers.

Heincke's Method

There are many ways of estimating Z from numbers in the catch. We shall describe one that dates as far back as 1913. Several other methods are described in Ricker (1975).

To introduce the method we start with a formula from common algebra:

$\overline{N}$, which is valid if a < 1

If a = exp(-Z) then for Z > 0 it follows that a < 1 and consequently the following expression is valid:

$\overline{B}$

When we multiply by the same constant, N(0), on both sides of the equal sign we get:

$\overline{B}$

Assuming total mortality, Z, to remain constant we have:

N(t) = N(0)*exp(-Z*t) and it can be written as

$$\frac{\overline{B}}{R}=\frac{Y}{R}*\frac{1}{F}$$

It can be shown that is equivalent to:

$$Z=-In\frac{N(1)+N(2)+N(3)+N(4)+\ldots\ldots}{N(0)+N(1)+N(2)+N(3)+N(4)\ldots\ldots}$$

or

$$Z=-In\left[\sum_{t=1}^{\infty}N(t)/\sum_{t=0}^{\infty}N(t)\right]$$

It should still hold if CPUE can be assumed proportional to the stock number so, replacing the N's by CPUE(t),can be used in the form:

$$Z=-In\frac{CPUE(1)+CPUE(2)+CPUE(age\,3\,and\,older)}{CPUE(0)+CPUE(1)+CPUE(2)+CPUE(age\,3\,and\,older)}$$

where the CPUE of the oldest age groups is lumped. The reason for lumping them is that it is often so that age groups can be relatively easily separated for the first two or three (or perhaps four) age groups, whereas it is difficult to separate the older age groups (Ricker, 1975). Also it is sometimes necessary to exclude the youngest age group when it is not fully exploited by the fishery. In that case CPUE(0) should not be included in the denominator and CPUE(l) not in the numerator.

Robson and Chapman's Method

Another method was introduced by Robson and Chapman, 1961 (in Ricker, 1975). They showed that the best estimate of Z from age composition data i.e. the numbers caught per age group, is:

$$Z=-ln\frac{A}{B+A-a}$$

where

$$A = N(1) + 2*N(2) + 3*N(3) + ...$$

$$B = N(0) + N(1) + N(2) + N(3) + ...$$

The variance of the survival rate, S = exp(-Z), is

$$VAR(S) = S*\{(S-(A-1))/(B+A-2)\}$$

Estimation of Z from a Linearized Catch Curve

A "*linearized catch curve*" is a graphical representation of the logarithms of numbers caught plotted against age. The methods to be described in this section are essentially based on age composition data and as such they have been applied to temperate fish stocks.

Because direct age reading is problematic for tropical fish stocks the methods have to be converted into versions which take length composition data as input. This is possible when growth parameters are available so that lengths can be converted into ages by using the inverse von Bertalanffy equation. The theoretical derivations start by using the age structured model, which will subsequently be converted into a corresponding length-structured model.

The Constant Parameter System

The linearized catch curve methods for estimating Z are based on the assumption of a "*constant parameter system*". This section is therefore used to explain this concept. Consider, as an example, a fish stock during the period 1971-1975 with a life span of five years after recruitment. For simplicity we assume only one cohort per year, recruiting on 1 January.

The figures, A are the numbers of survivors per cohort and per age group on 1 January. Because we consider many cohorts simultaneously, an index to specify the year of recruitment of a cohort, y, is needed. N(y,t) is the number of survivors attaining age t from the cohort recruited at the start of year y-t. In this particular case t only takes the values t = 0, 1, 2, 3 or 4.

Note that a specific cohort can be followed diagonally across the Table. Every year the survivors of that cohort advance to a new age group. Each column, on the other hand, contains the numbers of survivors of five different cohorts on 1 January of a particular year.

Table Illustration of the "constant parameter system" and the "variable parameter system". Note that the column of numbers in boxes consists of five different cohorts in the year 1971, whereas the diagonal of numbers in boxes consists entirely of the 1971 cohort.

A: VARIABLE PARAMETER SYSTEM					
age*/year	COHORTS, number of survivors				
	1971	1972	1973	1974	1975
t	N(71,t)	N(72,t)	N(73,t)	N(74,t)	N(75,t)
0	2105	1111	9560	1869	1236
1	2575	736	405	3817	618
2	155	1097	281	142	1193
3	102	58	298	109	59
4	6	38	18	138	29

B: CONSTANT PARAMETER SYSTEM					
age*/year	COHORTS, number of survivors				
	1971	1972	1973	1974	1975
t	N(71,t)	N(72,t)	N(73,t)	N(74,t)	N(75,t)
0	2560	2560	2560	2560	2560
1	942	942	942	942	942
2	346	346	346	346	346
3	127	127	127	127	127
4	47	47	47	47	47
*) Age from recruitment data					

Table contains parts of nine different cohorts, of which only one is given for its entire life span (the 1971 cohort). It appears that recruitment, N(y,0), where y = 1971, 1972, 1973, 1974 and 1975, has varied from year to year.

The 1971 cohort and the 1973 cohort are strong ones, whereas the 1972 cohort is a weak one ("strong" means that N(y,0) is considerably greater than the average recruitment). Now, suppose that the recruitment, N(y,0), remains constant every year, as is the case in Table. Assume further that also F and M remain constant, then the number of survivors and the numbers caught would be the same for all cohorts. An inspection of table B shows that in this case the number of survivors per year during the life span of a cohort equals the number of survivors within a particular year for each age group. Thus, in the case of constant recruitment it does not matter whether we consider one cohort over its entire life span, or if we consider all the different cohorts (the entire stock) in one particular year. (The major part of the theory of Beverton and Holt (1957) is based on the assumption of constant recruitment.)

We know for sure that the assumption of a constant parameter system is never strictly fulfilled in real life. However, we are often in a situation which forces us to make assumptions, which are known to be crude approximations to reality. It often happens that only by making such assumptions we are able to carry out an analysis of available data, and it is better to do a crude analysis than none at all. We speak of a "*pseudo-cohort*" when working with data from one year assuming these to resemble those of a cohort during its life span. Thus, the numbers 2105, 2575, 155, 102 and 6 in Table, form a pseudo-cohort, whereas the numbers 2105, 736, 281, 109 and 29 form a real cohort.

The Linearized Catch Curve Equation

Suppose that an estimate of the age composition of the catch during a year is available, i.e. pairs of observations:

age interval	*numbers caught*
t1-t2	C(t1,t2)
t2-t3	C(t2,t3)
t3-t4	C(t3,t4)
.	.
.	.

To develop a method to estimate Z from such data, the starting point is the catch equation:

$$C(t,t2) = N(t1) * \frac{F}{Z} * \left[1 - \exp\left(-Z * (t2 - t1)\right)\right]$$

Although the time interval from t1 to t2 is the first interval it will also be used in the following as the symbol for the general time interval, i.e. t2-t3, t3-t4, etc. is not linear in t1 (or t2), so some transformation is required to make the analysis a linear regression (cf. Section 2.6). The first step is to replace N(t1) using above equationwhich gives:

$$C(t,t2) = N(Tr) * \exp\left[-Z * (t1 - Tr)\right] * \frac{F}{Z} * \left[1 - \exp\left(-Z * (t2 - t1)\right)\right]$$

The second step is to take logarithms on both sides. After rearranging the terms we get:

$$In\,C(t1,t2) = InN(Tr) + In\frac{F}{Z} + Z * Tr - Z * t1 + In\left[1 - \exp\left(-Z * (t2 - t1)\right)\right]$$

At least t1 now appears in one linear term (namely: -Z*t1). The terms ln N(Tr) + ln (F/Z) + Z*Tr form a constant, as N(Tr), Tr, F and Z are assumed to remain constant. To simplify the notation, this constant is named "d", and the equation then reads:

$$\ln C(t1,t2) = d - Z*t1 + \ln[1 - \exp(-Z*(t2-t1))]$$

A linear expression in t1 has been obtained, except for the last term. Various methods suggest different ways of dealing with this term.

The first uses of this "*linearized catch curve equation*" date back to Edser (1908), Heincke (1913) and Baranov (1918) as reviewed in Ricker (1975). In the remaining part of above we shall discuss various special applications.

The linearized catch curve based on age composition data. One of the methods commonly applied in temperate waters to estimate total mortality is the "*linearized catch curve method with constant time intervals*", which has been reviewed in Beverton and Holt (1956), Chapman and Robson (1960), Robson and Chapman (1961) and Ricker (1975).

If t2-t1 in above equation remains constant (e.g. when age groups are considered, so that t2-t1 = t3-t2 = t4-t3 = ... = 1 year), the non-linear term becomes a constant, which can be included in the intercept term. Thus, if the constant g is defined:

$$g = d + \ln[1 - \exp(-Z^*(t2-t1))]$$

then Eq. 4.4.2.1 reads:

$$\ln C(t1,t2) = g - Z^*t \text{ or}$$

$$\ln C(t,t+\Delta t) = g - Z^*t$$

Above equation is the "*linearized catch curve equation with constant time intervals*", where the slope is -Z.

Example: Catch curve with constant time intervals, North Sea whiting

Table shows an example of a linearized catch curve analysis based on age composition data. Most often such examples come from temperate waters where direct age reading is possible and where spawning is confined to a short period, once per year.

This is also the case in this example which deals with whiting (*Merlangius merlangus*) caught in the North Sea.

Table gives the annual age composition of whiting catches during the period 1974-1980 (from ICES, 1981a). The figures in Table are numbers caught per year per age group (cohort), where C(y, t, t+1) = number caught in year y of age between t and t+1 years (in millions). For instance C(1976, 3, 4) = 159 For each year the total mortality,

Z has been calculated using Eq. 4.4.3.1. These are linear regressions where x = age, y = ln C(y,t,t+1) and b = -Z. Thus, recruitment and other parameters are assumed to have remained constant from 1974 to 1980.

year y	*1974*	*1975*	*1976*	*1977*	*1978*	*1979*	*1980*	*average*	
age t				*C(y,t,t+1)*				*1974-80*	
0	**599**	239	424	664	685	478	330	488	not used
1	678	**860**	431	1004	418	607	288	612	
2	1097	390	**1071**	532	335	464	323	601	
3	275	298	159	**269**	203	211	243	237	used in analysis
4	40	54	75	32	**69**	86	80	62.3	
5	6	9	13	18	8	**25**	31	15.7	
6	1	8	3	5	5	3	**8**	4.7	
7+	(6)	(0)	(1)	(0)	(1)	(1)	(1)	1.4	
Z	1.88	1.16	1.36	1.25	1.33	1.40	1.12	1.28	

However, we could also calculate an overall estimate of Z for the seven year period by using the average number caught per age group, as shown in the last column of Table. By doing so we (more or less) circumvent the problem of assuming parameters to remain constant, as e.g. variation in recruitment is levelled out in the averaging process. Note that in this particular case the assumption of constant recruitment is not crucial. Except for the year 1974 the annual estimates of Z do not deviate much from the overall estimate.

Figure shows the graph for the estimation of Z based on the average age composition. As appears from Table, data for the three first age groups have been excluded from all the regression analyses, since they do not conform to the straight line. The reason is that these age groups are not under full exploitation.

The fishing mortality is lower than that of the larger (older) fish because the fish of age groups 0 to 2 are so small that they escape through the meshes of the trawl-net. (The major part of the North Sea whiting are caught by trawls with codend mesh sizes from 70 mm to 110 mm.) Another reason for expecting a reduced fishing mortality on the small fish is that all of these may not yet have recruited to the fishing grounds and are still on the nursery grounds.

Thus total mortality, Z = M + F, does not remain constant for all age groups although this was one of the assumptions behind the linearized catch curve analysis. One way to circumvent the problem is to exclude those age groups we suspect not to be under full exploitation, as was done in this example. Which age groups to exclude is a somewhat subjective choice. Usually the points systematically deviating from the straight line are excluded. However, it is hard to give a general rule to decide when this deviation is sufficiently large to justify the exclusion of the point. As a rule of thumb: whenever you are in doubt about a point, exclude it. In the case of figure depicted, it is easy to make the choice since the five points used provide a very convincing fit to the model.

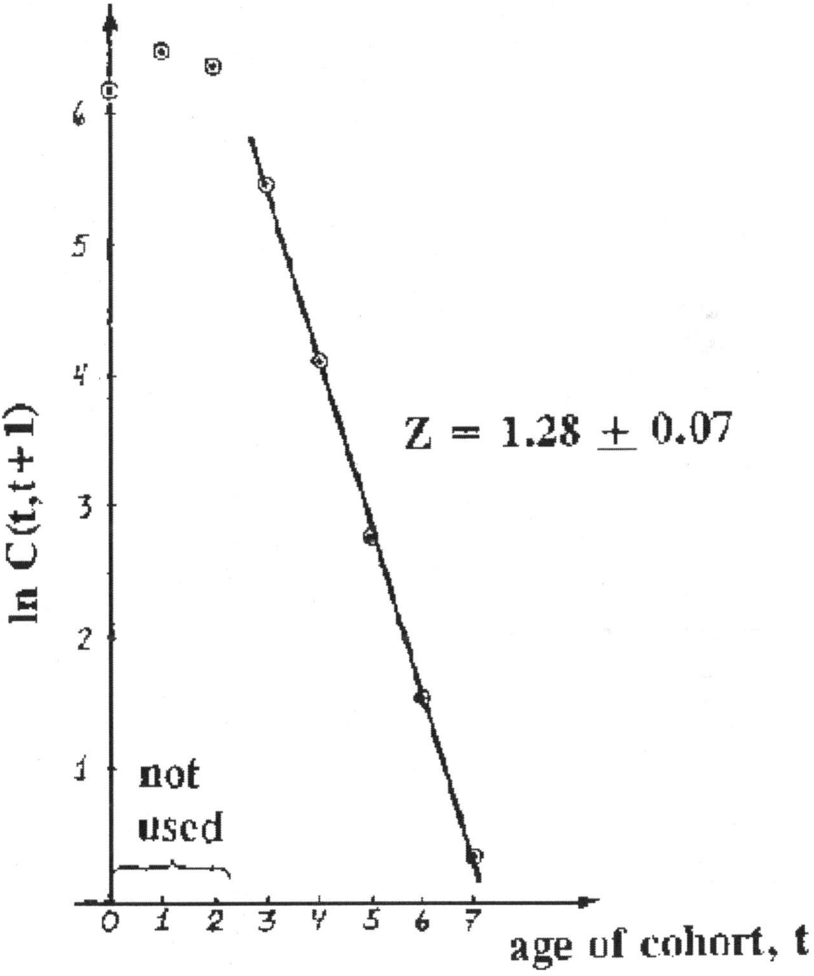

***Figure:** Linearized catch curve based on average age composition of North Sea whiting catches 1974-80 (ICES, 1981a)*

In the estimations of Z's for each year, age group 7+ has been excluded as well (indicated by brackets). The reason is that the number of specimens of age 7 or more (7+ groups) in each year has been estimated from small samples only. Because few fish survive to age 7 these estimates are subject to great uncertainty. However, for the overall estimate of Z the age group 7+ has been used, because in this case the samples cover seven years and the sample size is accordingly larger. Using the procedure described in Section 2.4 the 95% confidence limits of Z can be calculated:

$$Z = \text{-slope} = \text{-b} = 1.284$$

$$sy = 2.0316$$

$$sx = 1.5811$$

variance of the slope:

$$sb^2 = \frac{1}{n-2} * \left((sy/sx)^2 - b^2\right) = \frac{1}{5-2} * \left((2.0316/1.5811)^2 - 1.284^2\right) = 0.0005$$

$$sb = \sqrt{0.0005 = 0.0227}$$

$sb*t_{5\text{-}2} = 0.0227*3.18 = 0.07$, so that the confidence interval becomes

$$[1.21,\ 1.35] \text{ or: } Z = 1.28 \pm 0.07$$

With the data of table, the calculations could have been made on a cohort basis as well. Exercise deals with this aspect. The linearized catch curve based on age compositions with variable time intervals

Let us return to the general linearized catch curve model

$$\ln C(t1,t2) = d - Z*t1 + \ln[1 - \exp(-Z*(t2-t1))]$$

In case the time intervals t2-t1, t3-t2, t4-t3, ... are not constant the non-linear term, ln[1 - exp(-Z*(t2-t1))] takes different values for different values of the time interval. As will appear from Section 4.4.5 this case is often relevant to tropical fish stocks. Two alternative ways of getting rid of the non-linear term have been suggested.

Cumulated Catch Curve Equation

Jones and van Zalinge (1981) suggested to let t2 take a very high value (actually t2 = ∞ (infinity)). With t2 very large, the term exp(-Z*(t2-t1)) would be close to zero, consequently ln(1 - exp(-Z*(t2-t1))) = ln 1 = 0. Thus, if C(t,∞) designates all fish caught of age t and older, a linear relationship is achieved by

$$\ln C(t, \infty) = d - Z*t$$

C(t, ∞) is called the "*cumulated catch*", and the "*cumulated catch curve equation*".

Linearized Catch Curve Equation with Variable Time Intervals

Van Sickle (1977) suggested a different approach which was further developed by Pauly (1983a). It is based on the observation that for small values of x, (x < 1.0):

ln[1 - exp(-x)] = ln(x) - x/2 (approximately)

If Z*(t2-t1) is small it follows that approximately:

ln[1 - exp(-Z*(t2-t1))] = ln[Z*(t2-t1)] - Z*(t2-t1)/2

Inserting this into Eq. 4.4.2.1 and rearranging terms gives:

$$\text{In}\frac{C(t1,t2)}{t2-t1} = d + InZ - Z*t1 - Z*(t2-t1)/2$$

Because t2-t1 must be small, it is called Δ t. To simplify notation the constant c = d+ln Z is introduced. Then the equation reads:

$$\text{ln}\frac{C(t,t+\Delta t)}{\Delta 1} = c - Z*(t+\Delta t/2)$$

which is linear in the interval midpoint, (t+Δ t/2). Above quation is the so-called "*linearized catch curve equation with variable time intervals*". Note that only the slope has significance to the present analysis. The intercept is not used.

The name of the equation might lead to the misunderstanding that the method is applicable only when data for total catches are available. However can still be used even if only the percentage composition of the catches is known, because from these we get the absolute catches by multiplying all the percentages by a constant. A multiplication of all the values for C by a constant, K, will change the intercept, c, but not the slope, Z, since

$$\text{In}\frac{K*C(t,t+\Delta t)}{\Delta t} = c - Z*(t+\Delta t/2)$$

is equivalent to:

$$\text{In}\frac{(t,t+\Delta t)}{\Delta t} = c1 - Z*(t+\Delta t/2)$$

where c1 = c - ln K

Thus, if you have a sample from a fish population (for instance from a research survey or a market), but the total catch is unknown, you can use Eq. 4.4.4.2. Therefore the linearized catch curve analysis is often used for survey data, whereas cohort analysis is used for the analysis of total catch data. It should be noted that the example used in the next section to illustrate a linearized catch curve analysis based on length composition data is a sample of only 3816 fish, which was not raised to the total catch.

Beverton and Holt's Z-Equations

The first method dealt with in this section estimates Z from the mean length of the fish in the catch and the von Bertalanffy parameters K and L_∞. Thus, the data requirements are even less than those for the length-based linearized catch curve methods, for which the relative (not necessarily the absolute) size composition of the catch should be known. The derivation of Beverton and Holt's Z equations, however, is somewhat more complicated from a mathematical point of view. In the following we attempt to explain and justify the methods by aid of a numerical example, skipping the mathematics. It is finally shown that even L_∞ can be estimated together with an estimate of Z/K. Assume that samples of length compositions of catches have been collected from a fishery in 1960, 1970 and 1980, and that the results are those shown. The example used here is a hypothetical data set, that is, a data set for which the true parameters are known. The length groups represent only that part of the stock which is under full exploitation, the part for which Z is assumed to remain constant for all length groups. Z may of course be different for the three years in question. In addition, the von Bertalanffy parameters K and L_∞ must be known. For the example in Table, the parameters are:

$$K = 0.3 \text{ per year, } L_\infty = 100 \text{ cm}$$

The mean length L has a decreasing trend in time. Thus in 1980 fewer fish survived to a certain length compared to 1970 and 1960. If this change in mean length is caused only by a change in the fishing mortality on the stock it follows that:

$$(\text{F in 1960}) < (\text{F in 1970}) < (\text{F in 1980})$$

In other words, the lower the mean length in the catch, the higher is F. Under the assumption of a constant natural mortality, M, the same conclusion holds for the total mortality, Z = F + M.

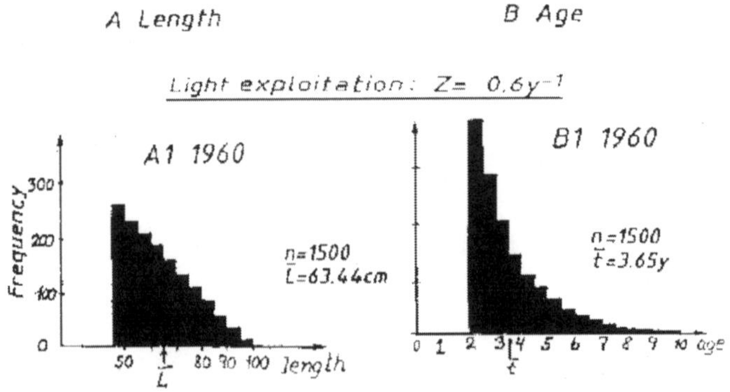

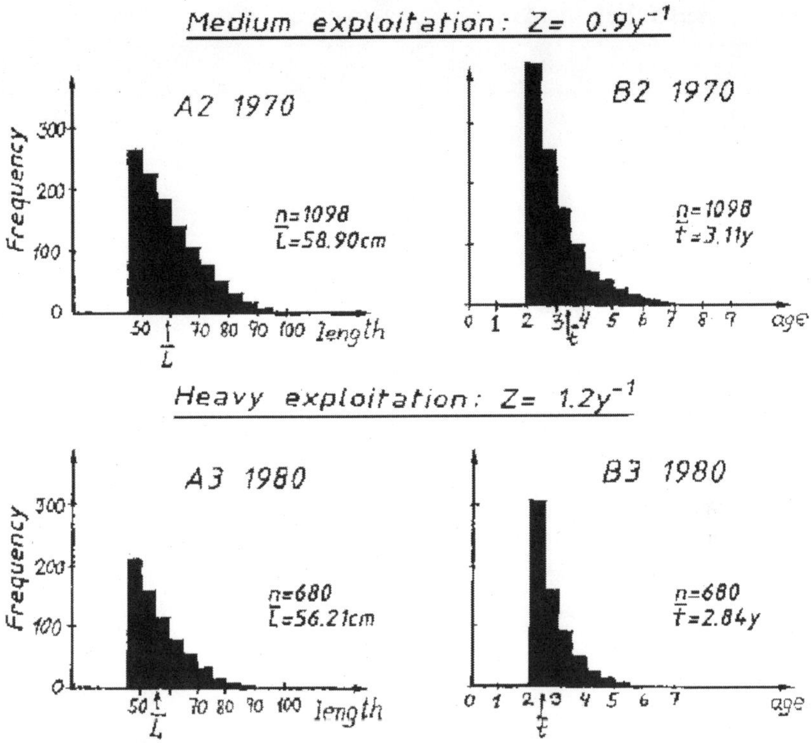

Figure: *Hypothetical example to illustrate Beverton and Holt's Z-equation.*
K = 0.3, L' = 45 cm and L_∞ = 100 cm

Beverton and Holt's Z-Equation based on Length Data

As was the case for the linearized catch curve method the length-based Beverton and Holt formula has an age-based parallel. It is mentioned here mainly because it illustrates a basic feature about the Beverton and Holt theory.

The age compositions corresponding to the length compositions in Tables and Figure. The same reasoning as was used for the mean length can be used for the mean age: the larger the fishing mortality the smaller is the mean age, $\bar{t}$, in the catch, or, the higher the mean age the smaller is F. This may be considered an implication of the functional relationship between age and length. However, the exact relationship between mean age and mean length is somewhat complicated. It should be noticed that the von Bertalanffy equation does not transform the mean age into the mean length, because the growth equation is not linear.

***Table:** Age composition of the samples given in Table*

Age Group	*1960*	*1970*	*1980*
2.0-2.5	390	399	308
2.5-3.0	289	255	169
3.0-3.5	214	162	92
3.5-4.0	150	103	52
4.0-4.5	118	66	28
4.5-5.0	97	42	15
5.0-5.5	64	27	8
5.5-6.0	48	17	6
6.0-6.5	35	11	2
6.5-7.0	26	7	-
7.0-7.5	19	4	-
7.5-8.0	14	3	-
8.0-8.5	11	2	-
8.5-9.0	8	-	-
9.0-9.5	6	-	-
9.5-10.0	4	-	-
10.0-10.5	3	-	-
10.5-11.0	2	-	-
11.0-11.5	2	-	-
sample size	1500	1098	680
Σ number*age	5475	3415	1931
mean age,	3.65	3.11	2.84

The age-based parallel to somewhat simpler. Beverton and Holt showed that:

$$Z = \frac{1}{\bar{t} - t'}$$

where $\bar{t}$ is the mean age of all fish of age t' and older, and where t' is "*some age for which all fish of that age and older are under full exploitation*". For the example age, t', corresponding to L' = 45 cm is

$$t'_{45} = t_0 - \frac{1}{K} * In\left(1 - L / L_{\infty}\right) = 0 - \frac{1}{0.3} * In\left(1 - 45/100\right) = 1.99 years$$

and

$$Z(1960) = \frac{1}{3.65 - 2.0} = 0.6 \, per\ year$$

$$Z(1970) = \frac{1}{3.11 - 2.0} = 0.9 \, per\ year$$

$$Z(1960) = \frac{1}{2.84 - 2.0} = 1.2 \, per\ year$$

Beverton and Holt's Z-equation based on length-at-first-capture.

A Plot of Z on Effort for Separate Estimates of F and M

The estimate of Z derived by the methods described in the foregoing section can sometimes be used to obtain estimates of F, the fishing mortality and M, the natural mortality. This is possible if pairs of estimates of Z and effort are available for a number of time periods covering a wide range of efforts.

Let Z(y) and f(y) be the total mortality and the effort in year y. The method (Paloheimo, 1958, 1961 and 1980) is based on Z = F + M, and the assumption that effort and fishing mortality are related in the simple manner of F(y) = q*f(y), where q is the catchability coefficient.

Z(y) = M + q*f(y)

With Z as the dependent variable and ϕ as the independent variable above equation becomes a linear regression with slope q and the natural mortality, M, as the intercept. The fishing mortality for the time period y, F(y), is derived from:

F(y) = Z(y) - M

Example: Estimation of M and q of a tropical fish

Table and Figure show an application of above equation to *Selaroides leptolepis* in the Gulf of Thailand (from Boonyubol and Hongskul, 1978). In this case Z is calculated, i.e. the observations are Lc, the 50% retention length, and $\overline{Lc}$, the mean length of all fish in the catch. Effort is given in units of millions of trawling hours per year. Figure shows the plot of Z on effort.

Using linear regression we find:

slope: q = 0.2532

intercept: M = 2.034 per year

Confidence limits are calculated as described in Section 2.4:

Table Data for estimation of M and q for *Selaroides leptolepis* in the Gulf of Thailand (from Boonyubol and Hongskul, 1978). K = 1.16 per year, Lc = 10.0 cm, L_∞ = 20.0 cm

year	effort a)	mean length	$Z = K * \frac{L\infty - \overline{Lc}}{\overline{Lc} - Lc}$ (Eq. 4.5.3.1)
y	f(y)	$\overline{Lc}$ cm	
1966	2.08	13.25	2.41
1967	2.08	13.01	2.69
1968	3.50	12.99	2.72
1969	3.60	13.07	2.62
1970	3.80	12.37	3.73
1971	-	- no data	-
1972	7.19	12.30	3.88
1973	9.94	12.01	4.61
1974	6.06	12.60	3.30
a) effort in units of millions of trawling hours			

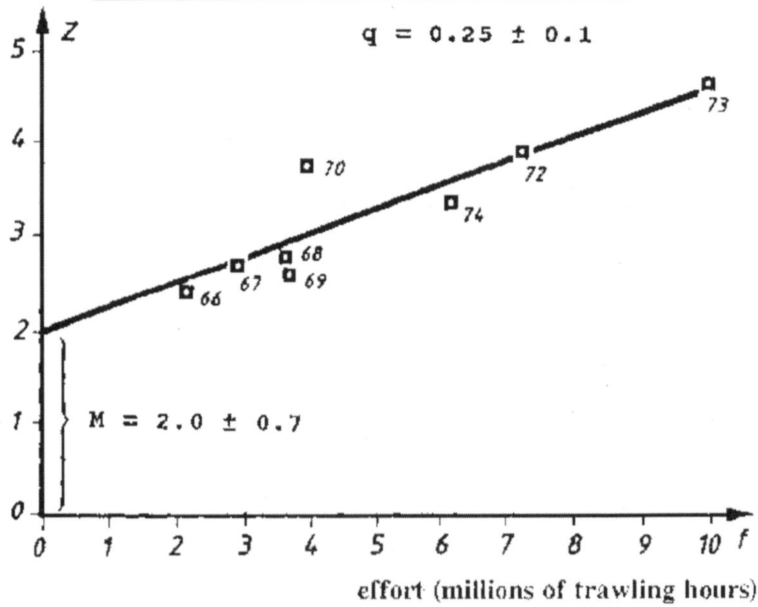

Figure: *Plot of total mortality, Z, on effort, f, for estimation of natural mortality, M, and catchability coefficient, q. Based on data from Table (from Boonyubol and Hongskul, 1978)*

$$sq^2 = \frac{1}{n-2} * \left[\left[\frac{sy}{sx}\right]^2 - q^2\right] = \frac{1}{6} * \left[\left[\frac{0.7724}{2.7423}\right]^2 - 0.2532^2\right] = 0.002533$$

sq = 0.0503

$$\overline{Ty} = \frac{1}{Z} + Tc..........(8.3.4)$$

sM = 0.2731

The 95% confidence limits for q and M are

q: [q - sq*t_{n-2}, q + sq*t_{n-2}] = [0.253 - 0.0503*2.45, 0.253 + 0.0503*2.45] = [0.13, 0.38]

M: [M - sM*t_{n-2}, M + sM*t_{n-2}] = [2.03 - 0.2731*2.45, 2.03 + 0.2731*2.45] = [1.36, 2.70]

The estimates of M and q obtained this way are not precise. For example, they are based on the assumption that effort is proportional to fishing mortality, i.e., that q is constant - an assumption which can always be questioned due to an increasing trend in fishing efficiency. One trawling hour in 1970 may not create the same fishing mortality as a trawling hour in 1980 because fishing vessels in 1980 were often better equipped (more efficient gears, larger engines, etc.) than the vessels in 1970.

If this is the case in the example of figure, the more recent observations should be moved to the right hand side to compensate for the increased efficiency. This would correspond to expressing the effort in units of 1966 trawling hours per year.

The effect on the regression analysis would be a smaller slope and a larger intercept, i.e. the estimate of M would be larger and that of q smaller.

Note also that the application of the method is only possible if there is a clear change in effort during the period considered. If effort (the independent variable) remains constant during the period we should not be able to fit a line to the points.

We shall continue the discussion of natural mortality, and two alternative simple methods for estimation of M will be suggested. All simple methods for estimation of natural mortality are questionable and it is recommended always to assess the reasonability of any estimate by comparing results of alternative methods and estimates of M for similar species.

Natural Mortality

One method of estimating the natural mortality coefficient, M, was introduced in Section 4.6. Many other procedures have been suggested in the literature. Most of them, perhaps all, rank no higher than "guesstimates" or "qualified" guesses. Some of these methods are widely used and others may be so in the future. They therefore merit discussion. Recall that the natural mortality is the mortality created by all other causes than fishing, e.g., predation including cannibalism, diseases, spawning stress, starvation, and old age.

Predation and starvation mortalities and several others are linked to the ambient ecosystem. The same species may have different natural mortality rates in different areas dependent on the density of predators and competitors whose abundance is influenced by fishing activities.

As direct measurements of M are often impossible to obtain, it has been attempted to identify quantities which can be assumed proportional to M and which are easier to measure (or estimate).

The von Bertalanffy curvature parameter, K, has been demonstrated to be linked to the longevity of the fish (Beverton and Holt, 1959) and longevity is related to mortality (Tanaka, 1960; Holt, 1965 and Saville, 1977). Other methods involving longevity are treated.

As a rough generalisation, fish species with a high K-value have a high M-value, and species with a low K-value have a low natural mortality. A slow growing species (low K) simply cannot bear high natural mortality - if it did it would soon become extinct. Beverton and Holt (1959) found that values of the ratio M/K mostly lie in the range of 1.5 to 2.5.

Natural mortality must also be linked to L_∞ or the maximum weight of the species, W_∞, since large fish have fewer predators than small fish. It has been suggested that M can be predicted from the body size for certain groups of animals (Taylor, 1960, for bivalves and Peterson and Wroblewski, 1984, for pelagic fish).

Rikhter and Efanov (1976) demonstrated that fish with a high natural mortality mature early in life, compensating for the high M by starting to reproduce earlier. It can also be mentioned that Gunderson and Dygert (1988) found a relationship of M with the ratio of gonad weight to somatic weight. This is reasonable because fish with a high mortality may compensate by producing more eggs.

Further, as most biological processes go faster at higher temperatures (of course within certain limits) one could imagine the natural mortality to be related to the environmental temperature. Pauly's empirical formula (Section 4.7.2) describes M as a function of K, L_∞ and T, the ambient temperature.

Natural Mortality and Longevity

Intuitively, we would consider longevity as something more closely related to mortality than K, L_∞ or ambient temperature. The basic concept of mortality coefficients was discussed. Alagaraja (1984) suggested another way of illustrating the concept of the mortality coefficient. He tentatively defined the natural life span of fish species (or the *longevity*) as the age at which 99% of a cohort had died if it had been exposed to natural mortality only (i.e. if Z = M). If Tm stands for longevity and M1% stands for the natural mortality corresponding to a 1% survival, then:

$$M1\% = -\ln(0.01)/Tm$$

Table shows a collection of M-values defined by the 99% life span as well as the corresponding 99.9% life span.

Table can be used for a first rough evaluation of a mortality estimate in conjunction with a modal progression analysis (in particular for tropical species). If, for example, from the modal progression analysis you find significant numbers (more than 1%) of three year old fish in the catch samples you should not accept a total mortality Z larger than 1.5. If you find very few (less than 1 per mille) you may accept a total mortality up to 2.3.

***Table:** M-values for various life spans*

Tm	*(months)*	*3*	*4*	*5*	*6*	*7*	*8*	*10*	*12*	
M1%	per year	18	14	11	9	8	6.9	5.5	4.6	
M0.1%	per year	28	21	17	14	12	10.4	8.3	6.9	
Tm	*(years)*	*2*	*3*	*4*	*5*	*6*	*8*	*10*	*15*	*20*
M1%	per year	2.3	1.5	1.2	0.92	0.77	0.58	0.46	0.31	0.23
M0.1%	per year	3.5	2.3	1.7	1.38	1.15	0.86	0.77	0.69	0.63

As longevity is usually as difficult to observe as the natural mortality the relationship between mortalities and life span does not provide any easier way to estimate M, but it presents the concepts in a way which may be easier to grasp. How easy it is to estimate the longevity of a certain species depends on how difficult it is to determine the age of the oldest specimens. If the age of the oldest specimens can be read from the hard parts and if the stock is unexploited it is relatively easy to get an estimate of Tm. The above aspects were discussed by Hoenig (1983), who also developed a model for estimation of Z from observations on longevity.

This model is basically the same as the Alagaraja (1984) model, but applies to Z rather than to M. Hoenig gave the following equation based on observations of Z and Tm and linear regression analysis:

$$\ln Z = a + b^{*}\ln Tm \text{ with}$$

a = 1.46 and b = -1.01 for fish (based on 84 stocks of 53 species) and

a = 1.23 and b = -0.832 for molluscs (based on 28 stocks of 13 species)

The above results were based mainly on data from unexploited stocks. Hoenig suggests Tm to be estimated from the mean age of "the oldest specimens" in a sample. This approach, however, is somewhat problematic as the estimate of Tm then depends on the sample size and on the proportion of the sample that is considered to represent "the oldest specimens".

Finally it should be emphasized that the usual assumption, that natural mortality remains constant for all age (size) groups within a species, is highly unrealistic for certain age (size) groups. Naturally, a small fish is exposed to a larger predation mortality than a big fish, simply because small animals have more predators than large animals.

This difference in predation mortality (which is perhaps the dominating source of natural mortality for small fish) may be quite big.

When estimating predation mortalities from multispecies cohort analysis combined with work on stomach contents, the ICES Multispecies Working Group (ICES, 1986; Gislason and Sparre, 1987), found a factor of about 10 between the mortalities of 0-group fish and 2-group fish in the North Sea.

Pauly's Empirical Formula

Pauly (1980b) made a regression analysis of M (per year) on K (per year), L_∞ (cm) and T (average annual temperature at the surface in degrees centigrade), based on data from 175 different fish stocks, and estimated the empirical linear relationship:

$$\ln M = -0.0152 - 0.279^{*}\ln L_\infty + 0.6543^{*}\ln K + 0.463^{*}\ln T$$

Table shows values of M (per year) calculated by "*Pauly's formula*" for various combinations of L_∞, K and T. It should be kept in mind that gives an estimate of M belonging to the category of "qualified guesses". Equation may be correct for the "average fish", but may be way off the mark for any particular fish stock. The formula indicates that:

Table Natural mortality (per year) calculated by Pauly's formula for various combinations of L_∞, K and T

	T = 5 C°				*T = 25 C°*			
$\bar{L}y = L_\infty\left[1 - \frac{Z*S}{Z+K}\right]$ ....(8.3.5)	***K 0.1***	***0.5***	***1.0***	***2.0***	***0.1***	***0.5***	***1.0***	***2.0***
10	0.24	0.7	1.1	1.7	0.51	1.5	2.3	3.6
80	0.14	0.38	0.6	1.0	0.29	0.8	1.3	2.0
200	0.10	0.30	0.47	0.7	0.22	0.6	1.0	1.6

1. Small fish have high natural mortalities
2. Fast growing species have high natural mortalities
3. The warmer the ambient water the higher the natural mortality

The implication is that any other aspect which may influence the natural mortality is considered "random noise around the regression line". Other aspects are, for example, the behaviour (schooling, pelagic/demersal), the reproduction physiology, the ecosystem (abundance of predators). Pauly (1983) suggests to account for schooling in above equation by multiplication by 0.8 so that for schooling species the estimate becomes 20% lower:

$$M = 0.8*\exp[-0.0152 - 0.279*\ln L_\infty + 0.6543*\ln K + 0.463*\ln T]$$

It should also be kept in mind that the quality of the input data to the regression analysis from which was derived, can be questioned. To arrive at Pauly needed "observations" of M, which as discussed in the foregoing are difficult to obtain. Any M value which was used for the estimation of above equation can be questioned. Therefore it is recommended only to use one decimal when presenting the result. For some species the result seems to be twice or half of what it should be. However, when no other information on M is available as is most often the case one should not hesitate to apply or the Rikhter and Efanov formula discussed. Usually, it is only for unexploited stocks (M = Z) that we are able to estimate M.

Note that Pauly's formula is based on data for fish stocks only, and that the result depends on how you measure the length of the fish (total length, fork length, standard length, etc.). As Pauly's equation gives only a rough estimate of M we can ignore this detail. Do not use Pauly's formula for crustaceans, molluscs or cephalopods or any other invertebrates, as the formula does not cover these groups.

Rikhter and Efanov's Formula

Beverton and Holt (1959) investigated a relationship between longevity, Tm, and the Lm/L_∞ ratio, where Lm is the length at first spawning. Holt (1962) noted that it was generally accepted that the Lm/L_∞ ratio was about 2/3 (average 0.64 with minimum and maximum values of 0.3 and 0.9). Continuing these investigations in comparative dynamics for high latitude stocks, Rikhter and Efanov (1976) showed a close association between M and Tm50% the age when 50% of the population is mature (also called "*the age of massive maturation*"):

$$M = 1.521/(Tm50\%^{0.720}) - 0.155 \text{ per year}$$

They also suggested that Tm50% should be equal to the "optimum age" defined as the age at which the biomass of a cohort is maximal.

Above equation gives these values for Tm50%:

Tm50%	*4 months*	*8 months*	*1 year*	*5 years*	*10 years*
M	3.2	1.9	1.4	0.3	0.1

6

Estimation of Gill Net Selection

Symmetrical Selection Curves

Gill nets are usually long rectangular nets where the upper edge, the head rope has floats while the foot rope has sinkers. Often gill nets (drifting and set nets) are in the form of gangs of nets with different mesh sizes. For further descriptions of gill nets. The selection properties of gill nets are reviewed in Hamley (1975). Discussions on gill net selectivity can be found in, for example, Baranov (1948); McCombie and Fry (1960), Gulland and Harding (1961), Regier and Robson (1966), Hamley and Regier (1973) and Jensen (1986).

Gill nets are "*passive gears*", i.e. the fish have to swim into the net to get caught. Theoretically, this implies that fish which move fast, have a larger probability of encounter with the gear than slow moving fish. Further it is known that large fish move faster than small fish of the same species. The swimming speed can be approximated by a constant times a power function of length:

$A*L^B$ where A and B are constants (Yates, 1983).

Rudstam, Magnuson and Tonn (1984) included swimming speed (with B = 0.8 for the cisco, *Coregonus artedii*, from Wisconsin, USA) into a model for gill net selection. They considered the selection as the product of two probabilities:

(selection) = (probability of encounter) *(probability of being caught given encounter)

We shall, however, only deal with the last factor, the probability of being caught given encounter.

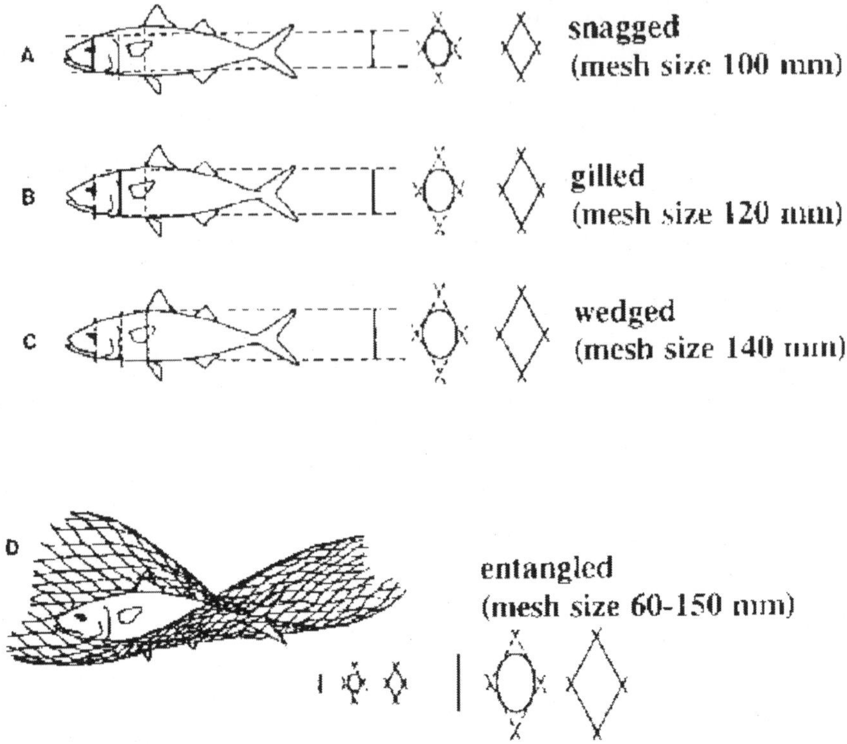

Figure: *Gill net selection (from Karlsen and Bjarnason, 1986)*

For simple gill nets the selection curve has (unlike trawl selection) a descending slope on the right-hand side. Small fish can pass through the meshes as was the case for trawl nets, but large fish may also avoid being caught in a gill net, because their heads are so large that they cannot be "gilled".

This is the simple theory behind gill net selection. The picture becomes somewhat more complicated when other ways by which the fish can get stuck in a gill net are also considered. Karlsen and Bjarnason (1986) distinguished four ways of getting caught:

a. Snagged, where the mesh is around the fish just behind the eye
b. Gilled, where the mesh is around the fish just behind the gill cover
c. Wedged, where the mesh is around the body as far as the dorsal fin
d. Entangled, where the fish is held in the net by teeth, maxillaries, fins or other projections, without necessarily penetrating the mesh.

Holt's Model for Two Mesh Sizes

For gilling and wedging Holt (1963) suggested a bell-shaped selection curve similar to the normal distribution. For these two types of getting stuck in a gill net we use the model:

$$S_L = \exp\left[-\frac{(L-Lm)^2}{2+S^2}\right]$$

where Lm is the "*optimum length for being caught*" and s is the standard deviation of the normal distribution. The factor "$n * dL / \left(s * \sqrt{2\pi}\right)$ " (which appears in the expression for a normal distribution is not used here. Omitting this factor, S_L becomes a fraction, i.e. $0 < S_L <= 1$.

Holt (1963) suggested an experiment to estimate Lm and s by using two gill nets with different mesh sizes, ma and mb. The two mesh sizes must be such that their selection curves overlap. The two nets are set to fish in the same area at the same time and the observations are numbers caught by length group. The assumptions behind this method are:

1. The optimum length Lm (the top of the bell-shaped selection curve) is proportional to the mesh size (Lm = SF*m, where SF is the selection factor.)
2. The two selection curves have the same standard deviation
3. The two gill nets have the same fishing power. This includes that when set, they must have the same length and height and be made of the same material.

Example: Selection curves for tilapia in Lake Victoria

Table shows an example of an experiment with two mesh sizes on *Tilapia esculenta* from Lake Victoria. The numbers caught, Ca, by length group for the smaller meshed gear (ma = 8.1 cm) and the corresponding numbers, Cb, for the larger meshed net (mb = 9.1 cm) are given. The parameters to be estimated by means of Holt's model are:

Lma: Optimum length for the smaller meshed net

Lmb: Optimum length for the larger meshed net

s: The common standard deviation

Input data for the analysis are the numbers caught by length group for each gear, Ca and Cb, and the two mesh sizes ma and mb. The mathematical derivations are lengthy and are bypassed here.

Step 1: Calculate the log ratios y = ln(Cb/Ca) for each length group. Only the lengths where the frequencies overlap can be used.

Step 2: Do a regression analysis of the log ratios (y = ln(Cb/Ca)) against the interval midpoint for fish length (x = L), and determine a and b:

$$\ln(Cb/Ca) = a + b*L$$

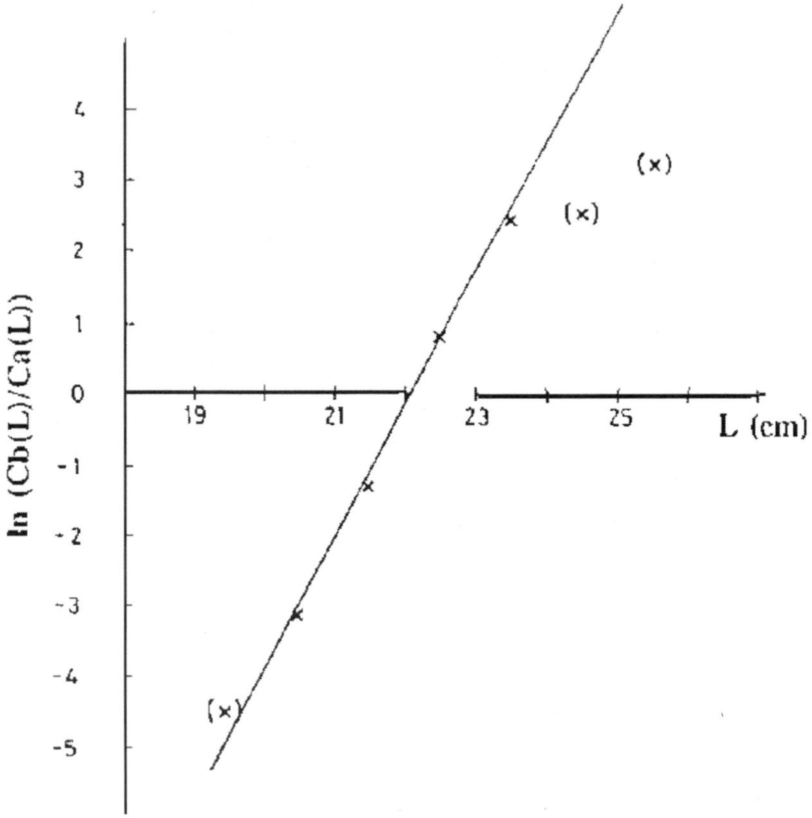

Figure: *Tilapia esculenta. The regression of ln(Cb(L)/Ca(L)) on fish length.*

Step 3: The results are finally obtained by inserting the values of a, b, ma and mb in the following expressions.

The selection factor is estimated from

$$SF = \frac{-2 * a}{b * (ma + mb)}$$

The optimum fish lengths for the small and large mesh size are respectively Lma = SF*ma and Lmb = SF*mb.

The common standard deviation s is determined by the variance

$$S^2 = \frac{-2 * a * (mb - ma)}{b^2 (ma + mb)} = SF * \frac{mb -}{b}$$

Step 4: Points on the selection curves are found by inserting values of L into above eqation:

$$\overline{N}$$

$$\overline{W}$$

Step 5: From these and the catches Ca(L) and Cb(L) an index of the numbers in the population is estimated for each mesh size:

Na(L) = Ca(L)/Sa(L)

Nb(L) = Cb(L)/Sb(L)

Table: *Estimation of gill net selection curves for* Tilapia esculenta, *Lake Victoria (data from Garrod, 1961)*

length interval midpoint L	*number caught*		$\overline{B}$	*selection*		*population estimates*	
(x)	*ma = 8.1* Ca_L	*mb = 9.1* Cb_L	*(y)*	Sa_L	Sb_L	Na_L	Nb_L
18.5	7	0	-	0.1333	0.0001	52	-
19.5	90	1	-4.500	0.5164	0.0036	174	(282)
20.5	199	9	-3.096	0.9583	0.0443	208	203
21.5	182	53	-1.234	0.8519	0.2611	214	203
22.5	119	290	0.891	0.3627	0.7373	328	393
23.5	29	357	2.510	0.0739	0.9970	392	358
24.5	17	225	2.583	0.0072	0.6458	(2492)	348
25.5	3	82	3.308	0.0003	0.2003	(8881)	409
26.5	0	19	-	0.0000	0.0304	-	(638)
27.5	0	10	-	0.0000	0.0021	-	(4721)
$\ln (Cb_L/Ca_L) = a + b*L$; a = -41.907; b = 1.894							
$\overline{B}$							
Lma = SF*ma = 20.84 cm							
Lmb = SF*mb = 23.41 cm							
$\overline{V}$							
s = 1.1655							

The indices Na and Nb are in principle the same for given L except for accidents of sampling. They should be calculated only for length groups for which there is a fair number of fish in the catch of the mesh size in question.

If the basic assumption that the selection curves are shaped like normal distributions is not fulfilled the estimation of the numbers in the population must be restricted to the part of the curve for which the assumption is approximately fulfilled. The catch curves are skew, probably an effect of entanglement: the big fish are over-represented in both. This is seen when plotting data for the regression analysis. The estimation is restricted to four points which are almost on a straight line and not based on very small numbers of fish. Na(L) and Nb(L) are in fair agreement for the four points used in the regression analysis.

For the big fish they are unreliable because the catch curves are skew. A larger number of mesh sizes is required to cover a wider range of the population size structure.

Model for Various Mesh Sizes

When there are n mesh sizes, all used together in nets of the same size, there will be n-1 estimates of the intercept, a, and the slope, b. Thus we have the results

[a(1),b(1)],[a(2),b(2)], ..., [a(n-1),b(n-1)]

corresponding to the mesh sizes:

[m(1),m(2)],[m(2),m(3)],....,[m(n-1),m(n)]

Each data set (each mesh size) was used twice to produce these estimates: set no. 2 was used first with set no. 1 to estimate a(1) and b(1) and again with set no. 3 to estimate a(2), b(2) and so on. This introduces a correlation between consecutive pairs a(i), b(i) whose effect is difficult to ascertain, but may be important when data for only a few mesh sizes are available. Bearing this in mind we continue by rearranging give a straight line through the origin:

$$-2*a(i)/b(i) = SF*[m(i)+m(i+1)],\ i = 1,2,\ ...,\ n-1$$

from which the common selection factor, SF, is estimated as the slope, b.

With y(i) = -2*a(i)/b(i) and x(i) = m(i)+m(i+1) we have:

$$SF = S\ [x(i)*y(i)]/S\ x(i)^2$$

The common variance is estimated as the mean of the individual estimates for each consecutive pair of mesh sizes:

$$\overline{B} = \sum_{t=1}^{12}\left(\overline{B}(t)\Delta t\right) / \sum_{t=1}^{12}\Delta t$$

The optimum length for each mesh size i is obtained by

$$Lm(i) = SF*m(i)$$

Subsequently, points on the selection curves Si_L are calculated for each length group. Then these values are summed for each length group and normalized to a maximum of one, in order to get a composite selection curve. This is done by dividing the values of the sums by the highest value amongst the sums.

A flat curve over a wide range of lengths indicates that all length groups within this range were sampled evenly such that the sum of the catches for each length group Ci_L can be used directly as an index of population numbers.

If the composite selection curve has conspicuous summits and valleys the population numbers must be estimated by dividing the catch of each length class by the value (≤ 1) on the composite selection curve. There may be length groups for which the population index cannot be calculated.

Example: Estimation of combined selection of several gillnets for a cichlid in Lake Kariba

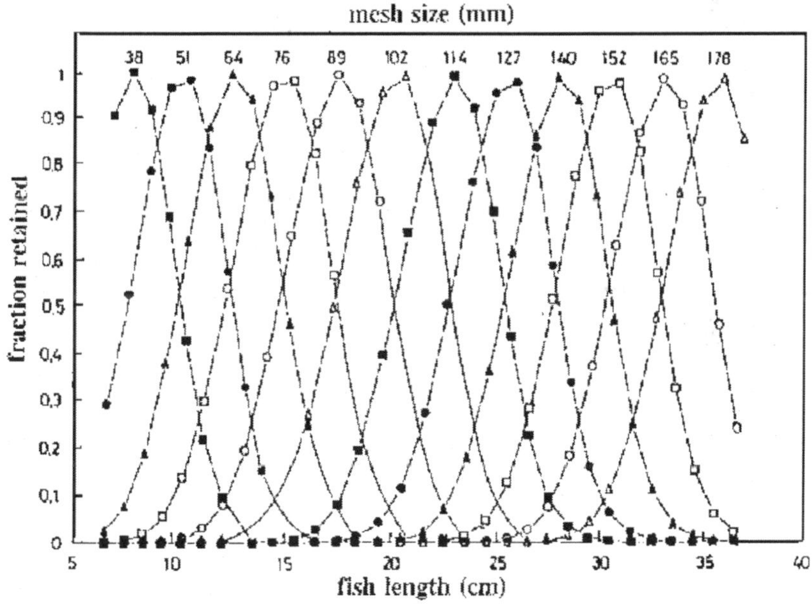

Figure: *Selection curves for the cichlid* Serranochromis codringtoni *in Lake Kariba, southeast Africa. Gill nets of 12 mesh sizes. Redrawn from Zambia/Zimbabwe/SADC Fisheries Project, Project Report No 26, 1993*

Figures show selection curves for a cichlid in Lake Kariba. In an experiment 21000 specimens of *Serranochromis codringtoni* were caught in a gang of 12 mesh sizes of gill nets ranging from 38-178 mm stretched mesh. The curves were estimated as described above. After summation and normalisation to one it is seen that the resultant composite selection curve is flat-topped for fish lengths of 11 to 31 cm. Thus, the total catch in the nets of 12 mesh sizes gave an almost true index of the size composition in the population for which was estimated $L_\infty = 27$ cm.

The increase in the left-hand side of the broad selection curve resembles a trawl selection curve. The decrease to the right will not be observed because there was no fish that big. It should be remembered that the data was from experimental fishery.

The commercial fishery in the lake is mainly with gill nets of one mesh size. The fishing mortality is therefore not expected to be the same from a certain size and up, as in a commercial trawl fishery.

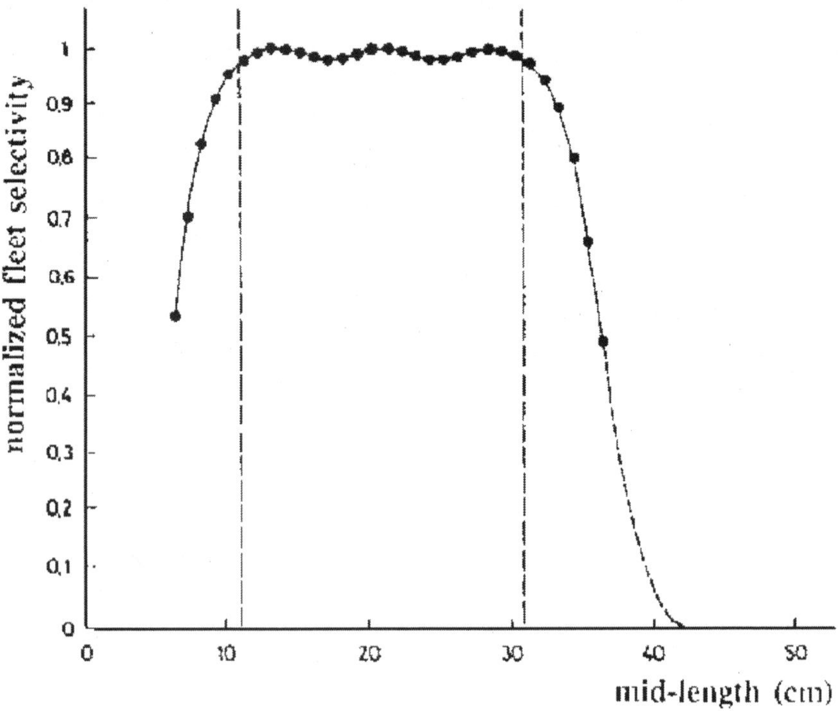

Figure: *The selection curves of Figure given earlier summed and adjusted to a maximum value of one. Redrawn from the above mentioned report*

Hanging Ratio

Gill net selection is known to depend on a variety of factors besides mesh size: net construction, visibility and stretchability of the net, net material and the shape and behaviour of the fish (Hamley, 1975). Entangling more than wedging and gilling is affected by net construction. The probability of a fish being entangled is believed to depend on the so-called "hanging ratio" or "hanging coefficient" which is defined (FAO, 1978b) as:

$$\overline{B} = \frac{\sum \overline{B}(t)}{12} = \frac{37758}{12} = 3146.5\ :$$

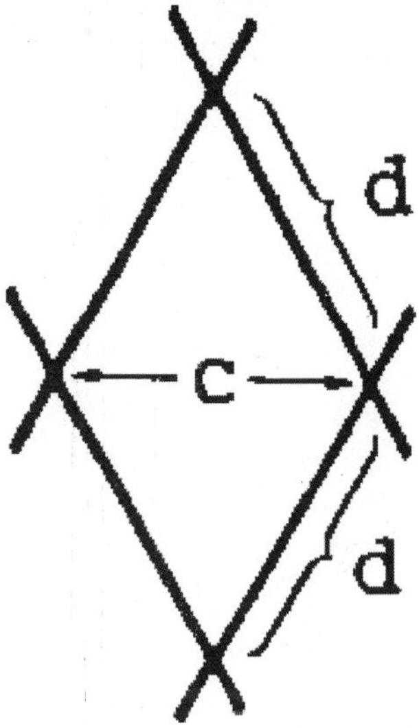

$$hanging\ ratio = \frac{c}{2 * d}$$

Thus, for a square mesh (d = c/√2) we have the hanging ratio √2/2 = 0.707 which represents the maximum opening. Hanging ratios are usually in the range of 0.2 to 0.7. The smaller the hanging ratio the larger the probability of entangling. This is demonstrated by Riedel (1963), who reported catches of *Tilapia mossambica* with 10 cm mesh gill nets with three different hanging ratios:

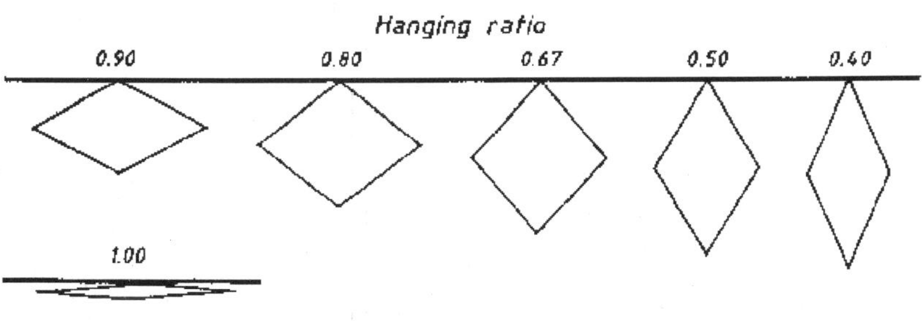

Figure: *Mesh shapes with various hanging ratios for gill nets (from FAO, 1978b)*

Hanging Ratio	*Average Number Caught per Day*	*Percent Entangled*	*Size Range of 95% of Catch*
0.707	9.3	0	18-23 cm
0.36	29.5	24	13-23 cm
0.24	81.0	80	8-22 cm

The Product of Two Logistic Curves

If entangling is an important factor, the normal distribution method described above is not suitable. One way of estimating the selection curve would be to compare the catches of the gill net to a non-selective gear, e.g. to trawl catches. The catches from the non-selective gear would then play the same role as the total combined catch in the cover and the codend of a trawl when operated as described in this section. The same procedure as used in Table, can be applied. In this case we need a non-symmetrical selection curve of the type shown earlier. A mathematical expression for this type of curve can be obtained by multiplying two logistic curves (Hoydal *et al.*, 1982).

The ascending part of the curve is given by the usual logistic curve, it reflects the probability of being gilled or wedged. We call it "SL" where "L" stands for "left-hand side of the selection curve":

$$SL_L = 1/(1 + \exp(S1 - S2*L))$$

This type of selection is the dominating one up to length A. For lengths larger than B the selection is the combined effect of gilling, wedging and entangling. This part of the curve is modelled by a "*reversed logistic curve*". We call it "SR" where "R" stands for "right-hand side of the selection curve":

$$SR_L = 1/(1 + \exp(D1 - D2*L))$$

The parameters in the SR-function, Dl and D2, are negative numbers, whereas the parameters in the SL-function, S1 and S2, are positive numbers. The lengths corresponding to the 50% and 75% de-selection, D50% and D75%, are related to D1 and D2 by the same mathematical expressions as those used for S1, S2, L50% and L75%:

$$D50\% = D1/D2 \text{ and } D75\% = (D1 + \ln 3)/D2$$

$$D1 = D50\%*\ln(3)/(D75\%-D50\%) \text{ and } D2 = D1/D50\%$$

By multiplying the ascending curve, SL, and the descending curve, SR, we obtain the desired type of curve, S. When SL is ascending SR is "neutral" (i.e. approximately equal to 1.0), and when SR is descending SL is "neutral":

$$S_L = SL_L * SR_L = \frac{1}{1+\exp(S2*L)} * \frac{1}{1+\exp(D1-D2*L)}$$

The expression for S_L is supposed to take the maximum value, 1.0, for at least one L-value. Therefore, the expression should be normalized so that the maximum value.

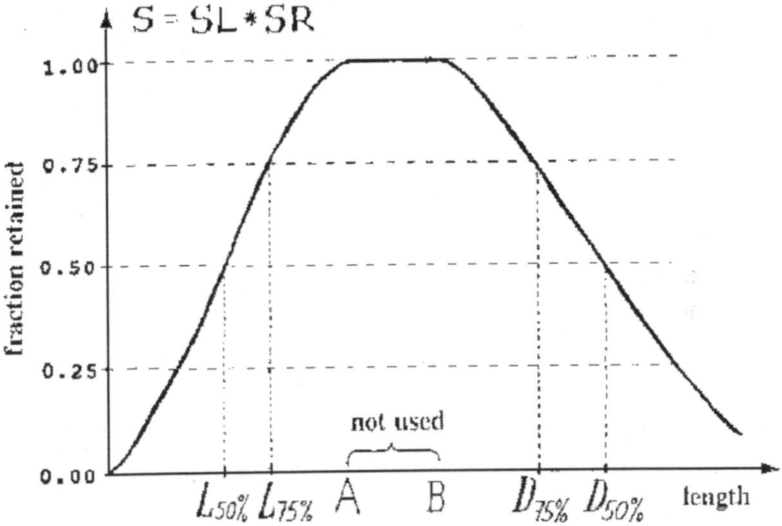

Figure: *Asymmetric selection curve*

In practice this is obtained in the following way. Let S(i) be the point on the selection curve representing length class no. i and let MAX {S(j)} designate the maximum value of S(j) among all length classes. We normalize by replacing the value of S(i) by the value:

$$\frac{S(i)}{\underset{j}{MAX}(S(i))}.$$

The parameters can be estimated in the same way as the parameters of the trawl selection ogive (cf. Section 6.1). To estimate S1 and S2 we use only the length classes below A and perform the regression analysis:

$$\ln(1/S - 1) = S1 - S2*L$$

where the dependent variable, y = ln(1/S - 1), is derived from the comparison with the non-selective gear:

$$S(i) = \frac{Cg(i)/Cn(i)}{MAX\{Cg(j)/Cn(j)\}}$$

where

$$\frac{Cg(i)}{Cn(i)} = \frac{no.of\ length\ class\ i\ fish\ caught\ in\ gill\ net}{no.of\ length\ class\ i\ fish\ caught\ in\ non-selective\ gear}$$

The denominator in Eq. 6.2.2.5, "MAX {Cg(j)/Cn(j)}", is the maximum value of the ratio Cg/Cn among all length groups with non-zero values of Cg and Cn. Thus, S(i) as defined by Eq. 6.2.2.5 takes values between 0 and 1 (including 1).

The assumption that the numbers Cn are caught by a gear which is non-selective for all those length classes which are caught by the gill net. If this is not the case should be replaced by:

$$S(i) = \frac{\left[Cg(i)/Cn(i)*Sn(i)\right]}{MAX\left\{\left(Cg(j)/Cn(j)\right)*Sn(j)\right\}}$$

where Sn(i) is the selection curve for the "other gear".

The descending (right-hand side) of the selection curve is estimated using the same calculation procedure and the corresponding observations as those used for the left-hand side. Thus we use the data for the length classes above point B and perform the linear regression analysis:

$$\ln(1/S - 1) = D1 - D2*L$$

The method based on the product of two logistic curves is a generalisation which includes the trawl selection ogive and the symmetric curve as special cases. Assigning the values $D1 = -\infty$ and $D2 = 0$ makes the factor $1/(1 + \exp(D1 - D2*L))$ in take the value 1 for all values of L and then equals. If the curve is symmetrical we estimate parameters so that:

$$L50\% + D50\% = L75\% + D75\%$$

but we do not have to make assumptions beforehand.

The symmetrical curve created by the product of two logistic curves may not be the same as the normal distribution curve. The product of two logistic curves may take the maximum value (1) for a range of values. It may also, however, come very close to the normal distribution as shown by an example in figures.

Thus, if data for a non-selective gear (or a gear with a known selection curve) are available, there is really no need to use the traditional model, the normal distribution, since the same curve can be obtained as a special case of the logistic product function. Moreover, the latter is versatile and easy to handle from a computation point of view. Using the product of two logistic curves we are not forced to make the questionable assumption that the selection curves are normally distributed with a common standard deviation.

Finally, it should be mentioned that data collected from gill net catches are difficult to use for the estimation of growth parameters or mortality rates. This is illustrated in figure earlier as a hypothetical example. We consider a length-frequency sample representing the population (double line) i.e. a sample from a non-selective gear and a sample representing the gill net catch (shaded line).

The population contains four components or cohorts (normal distributions), whereas the gill net sample appears to contain only one component with mean value Lm, which does not coincide with any of the four cohort mean lengths (L0, L1, L2 and L3).

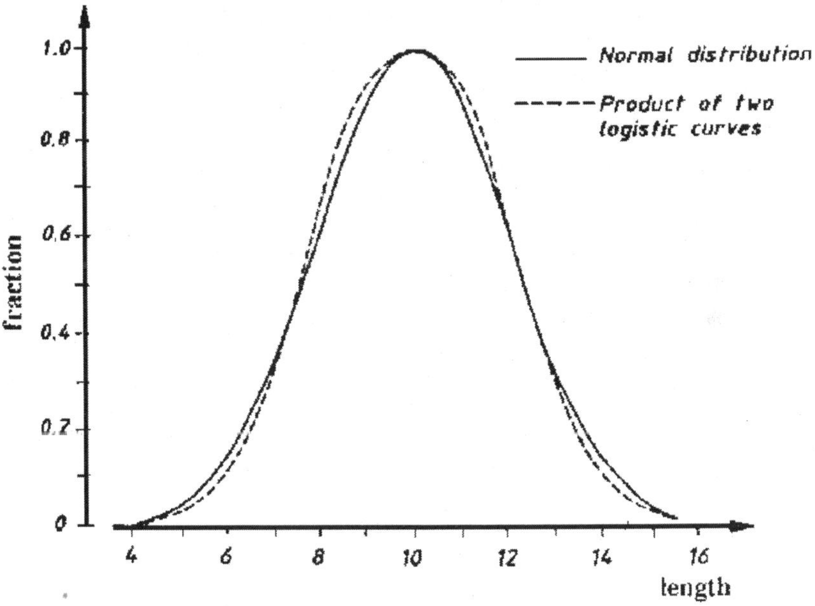

Figure: *product of two logistic curves which is nearly identical to a normal distribution. The parameters are:*

Normal distribution: Lm = 10 and s = 2

Product of two logistic curves:

L50% = 7.645

L75% = 8.483

D50% = 2*Lm - L50%

D75% = 2*Lm - L75%

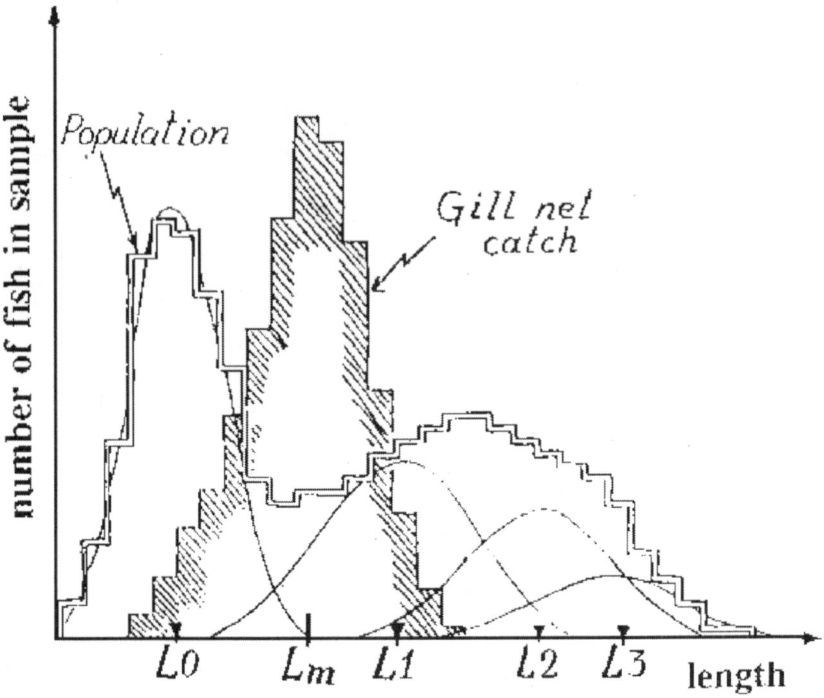

Figure: *Hypothetical example to illustrate bias problems when using gill net data for estimation of growth parameters and mortality rates*

Thus, the gill net sample does not give any information which can be used for the separation of cohorts and the estimation of length-at-age. Also samples collected during a whole year would all give more or less the same picture. Summing a time series over a year would therefore give a curve not very different from the curve for the gill net catch. Using the descending slope of this curve to estimate total mortality from the length-based catch curve analysis would lead to an overestimate of Z. Before using any gill net data for estimation of growth parameters or mortality rates, they should be critically examined. The result of such an examination may be either that the data cannot be used at all or that they can be used only after having been adjusted for gear selection.

Discussion of Selection by Other Gears

The previous two sections dealt with the selectivity of trawl nets (active gears) and gill nets (passive gears). The scientific literature on selectivity has been mainly concerned with these gears, partly because it is relatively easy to conduct experiments for the estimation of their selection curves. Other types of gear are also more or less selective and the pattern of selection can usually be changed by suitable adjustments of the gear.

The gear selection model based on the logistic curve or the model based on the product of two logistic curves are believed versatile enough to describe the selection curve of any gear.

Below is a brief discussion of the selection properties of two passive gear types: hooks and traps, and of an active gear type: seine nets.

Much more appears to be known about trawl and gill net selection than about hook fisheries. For hook-and-line fisheries some authors report that the selection curve is bell-shaped, of the gill net type, dependent on hook size, whereas others find a trawl type selection.

The idea behind the use of the bell-shaped selection curve is that small fish cannot take a large hook in their mouth and that large fish are not held securely by small hooks.

A discussion of hook selection can be found in Ralston (1982), who observed the trawl type of selection for the Hawaiian deep-sea hand line fishery. From an experiment where four hook sizes were used for catching snappers and groupers he found that small hooks were nearly as efficient as large hooks at catching large fish.

For the ascending left-hand selection curve (the small fish) he found a sigmoid curve. Pope *et al.* (1975) suggested using the gill net type of selection for hook fisheries, although they quote a number of works which show the trawl type of selection.

Pope *et al.* (1975) suggested the trawl type of selection for traps arguing that traps behave like codends in retaining fish. Munro (1974, 1983) discusses selectivity and other aspects of the operation of portable Antillean fish traps.

These 3 m long traps are made of chicken wire-netting with two funnels whose bottoms are the entrance openings. They are used for catching coral reef fish. Munro (1974) developed a model for the catchability of traps as a function of the soak time (the time the trap is left on the fishing ground).

Trap selectivity is complicated because it relies on the fish to move actively into the trap. For small traps (as the Antillean trap mentioned) only one specimen of a territorial fish species can be expected to be caught, but the trapped fish may have been replaced by another before the trap is pulled. As the other extreme we have the hunters, like the jacks (Carangidae) in a coral reef fishery which cover a lot of ground in a short time and therefore become over-represented in trap catches.

If a large predator is caught in the trap it may keep potential prey fish from entering the trap. If prey species are already in the trap they may act as live-bait and attract the large predators, which in turn may eat the prey before the trap is pulled.

Trap catches depend on the duration of the soak (Munro, 1974). There is always a chance that a trapped fish finds the entrance opening and leaves through it. There are considerable differences between species.

Some species leave the trap with great ease (Munro, 1983). Thus, trap selectivity may not only be a function of the mesh size used in the trap. For example, the size of the entrance opening and the soak time may be of importance.

The species composition on the ground where the trap is placed also may influence the selectivity. For the escapement through the meshes, however, it appears reasonable to assume the trawl selection type of ogive. When considering the average of a large number of trap catches some of the above mentioned complications may disappear, i.e. they may turn out to be "random noise" around the selection curve.

In principle a seine net should work like a trawl as far as selection is concerned. However, it is more difficult to deal with the seines because this type of gear is usually used to catch schooling species, such as sardines, mackerels and tunas. These species have a tendency to form schools consisting of fish of the same size. Thus, we should consider a school to be a sampling unit (instead of an individual fish).

Other Aspects of Gear Selectivity

Knife-Edge Selection

Curve A has a selection range of 3 cm and curve B, the fat vertical line, has a selection range of 0 cm. Curve B is a so-called "*knife-edge selection curve*" (Beverton and Holt, 1957). Knife-edge selection should be considered a hypothetical model since it will never describe a real situation. However, knife-edge selection is often used as an approximation to the selection ogive.

For lengths below L50% the numbers selected are under-estimated and for lengths above L50% the numbers are overestimated. These two sources of bias have opposite signs and as the two areas "a" and "b" are the same size they balance out. However, the fish of area "a" will weigh more than those of area "b", since the weight of a fish corresponds to the cube of the length.

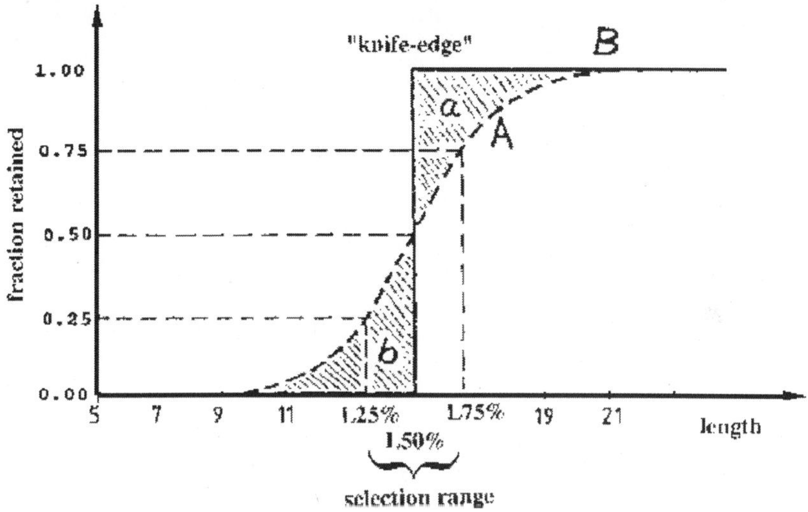

Figure: *Trawl selection curve as a function of body length. An illustration of the concepts of knife-edge selection and selection range*

Recruitment and Selectivity

The recruitment of fish to the fishing area, that is when they move away from the nursery or spawning areas to the fishing grounds, is also size dependent, in the same way as a trawl selection ogive. This means that every size of fish will not be fully represented at the fishing grounds and, thus, when there is a fishery for the size ranges which are not yet fully recruited the probability that a fish is retained by the fishing gear is in fact the product of two probabilities:

1. The probability that the fish is present on (has recruited to) the fishing ground.
2. The probability that the fish is retained by the meshes once it has entered the gear.

Fig. illustrates these points. Curve R is the "*recruitment curve*", curve G is the "*gear selection curve*", and curve S is the "*resultant curve*".

The probability that a certain size of fish will be caught is the product of the probabilities of recruitment and selection. The probability can therefore be described by a "*resultant curve*", S, where S = R*G. The L50% for the three curves, R, G and S, are different as indicated in figure depicted. The probability of capture of a fish of length Q is the product of probability A, related to the recruitment curve, and probability B related to the gear selection curve, and the result is probability C. In this example A*B = C or R(Q)*G(Q) = S(Q) or 0.62*0.42 = 0.26. At length X practically all fish have recruited to the fishing grounds while some are still small enough to escape through the meshes. At length Y mesh retention is complete and no fish escape. When the meshes are so large that there is no overlap of the recruitment curve with the selection curve we can ignore recruitment. The resultant curve is then determined by selection only.

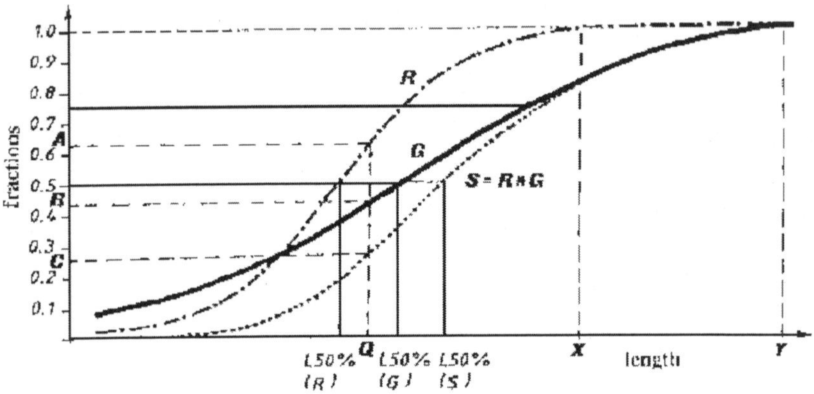

Figure: *Curves representing: recruitment to the fishing grounds (R), gear selection (G), and the resultant (S = R*G)*

Selectivity as a Function of Age

So far, selectivity has been considered a function of length. Another example of a selection curve S_L is given. The values of S1, S2, L50% and L75% corresponding to this curve are as follows:

S1 = 15*ln(3)/(18-15) = 5.4930 and S2 = ln(3)/(18-15) = 0.3662

L50% = 5.4930/0.3662 = 15 cm and

L75% = (5.4930+ln 3)/0.3662 = 18 cm

The selection range is 2*(18-15) = 6 cm

Using the von Bertalanffy growth equation we can express length as a function of age, and express S as a function of age, t:

$$S_t \frac{1}{1+\exp\left[S1 - S2 * L_\infty * \left(1-\exp\left(-K*\left(i-t_0\right)\right)\right)\right]}$$

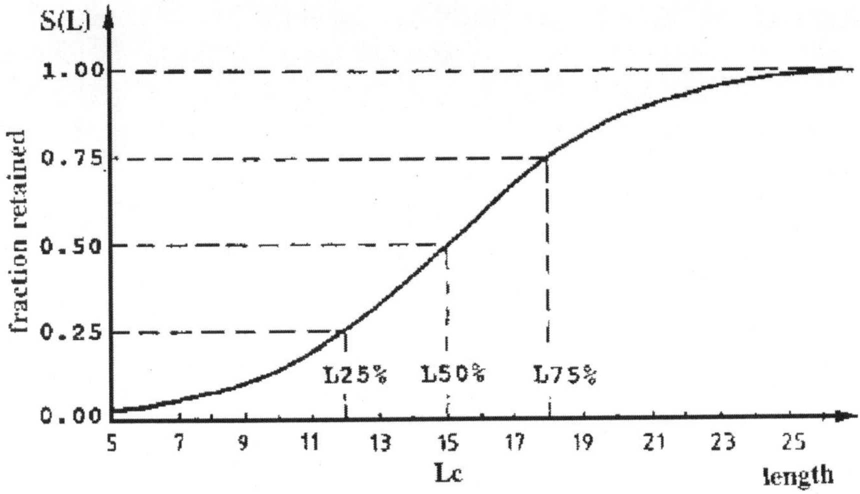

***Figure:** The selection curve referred to in the text*

The next, age-based expression is equivalent to which is length-based.

$$S_t \frac{1}{1+\exp(T1 - T2 * t)}$$

and which can be rewritten in linear form as:

$$\ln(1/S_t - 1) = T1 - T2*t$$

where:

T1 = t50%*ln(3)/(t75%-t50%)

T2 = ln(3)/(t75%-t50%) = T1/t50%

The formulas for t50% and t75% are:

t50% = T1/T2

t75% = (T1 + ln 3)/T2

Figure shows the two age-based selection ogives. Curve A is based on the exact transformation according to above equation, while curve B is the approximate selection ogive and the related equations. It can be observed that both curves are almost identical.

It is also possible to express t50% and t75% in lengths and vice versa to express L50% and L75% in ages, using the following formulas:

$$t50\% = t_0 - (1/K)*\ln(1 - L50\%/L_\infty)$$

$$t75\% = t_0 - (1/K)*\ln(1 - L75\%/L_\infty)$$

and

$$L50\% = L_\infty*[1 - \exp(K*(t_0 - t50\%))]$$

$$L75\% = L_\infty*[1 - \exp(K*(t_0 - t75\%))]$$

Assume that $L_\infty = 50$ cm, $t_0 = 0$ years and K = 0.3 per year for the fish stock associated with the length-based selection ogive in Figure, then:

$$t50\% = 0 - (1/0.3)*\ln(1 - 15/50) = 1.1889$$

$$t75\% = 0 - (1/0.3)*\ln(1 - 18/50) = 1.4876$$

$$T1 = 4.3727 \text{ and } T2 = 3.6779$$

The corresponding L50% and L75% are:

$$L50\% = 50*[1 - \exp(0.3*(0 - 1.1889))] = 15.0 \text{ cm}$$

$$L75\% = 50*[1 - \exp(0.3*(0 - 1.4876))] = 18.0 \text{ cm}$$

which are the same as the results obtained for the length-based selection ogive.

Estimation of the Resultant Ogive from a Catch Curve

When using a linearized catch curve to estimate mortality, it is usually necessary to discard the left-hand side of the curve because the juvenile fish are not fully exploited or not fully recruited.

A conceptually simple way to estimate how many fish are missing at each age is to extrapolate on the straight line from which the total mortality coefficient Z is estimated, in order to find the number of juveniles there "ought to be".

The differences between the "expected" numbers and the actual numbers should give the ogive resulting from the combined effect of recruitment and mesh selection. As shown below, the calculations are easily performed.

The problem is that an important and probably unrealistic assumption is made, namely that the total mortality rate, Z = F + M, is the same for all ages. F alone is not constant because it must be smaller in the mesh selection phase.

M, on the other hand, is likely to be higher for small fish than for the adults. It is therefore possible that Z remains approximately constant although so far, nobody has shown it to be. Nevertheless, the method has achieved considerable popularity and is therefore mentioned here.

Example: Estimation of the resultant selection ogive from a catch curve, hypothetical data

To explain this method (Pauly, 1984a) the example of Table is used. Columns A-E contain the input data and calculations for a length-converted catch curve analysis. In this case we calculate the total mortality Z = 1.0 per year from the growth parameters, $L_\infty = 50$ cm, K = 0.3 per year. The result of the regression analysis is:

$$In\frac{C}{\Delta t} = 9.208 - 1.0 * t$$

In contrast to the example discussed in Section 4.4.5 we now have a use for the intercept (a = 9.208).

Under the assumption of constant mortality we expect the values of ln(C/D t) to be on the regression line, ln(C/D t) = a-Z*t. Thus, the hypothetical true frequency the total population numbers in the sea, CT, is expected to fulfill the equation:

ln(CT/D t) = a - Z*t

The idea behind this method is that the number in the sea is proportional to the number caught, i.e.

$$\frac{C}{CT} = \frac{\textit{the number in the catch}}{\textit{total population number in the sea}}$$

Let t1 be the age corresponding to the first length group which is supposed to be fully represented in the catch and therefore used in the catch curve regression (in the case of table, we have t1 = 2.180). For ages above t1, CT_t should be approximately equal to the observed frequencies, since the probability of capture is 1, because selection and recruitment are supposed to have finished before that age. For the ages below t1 we expect that the population in the sea is higher than that represented in the catch, i.e.:

$\ln(CT_t/D\ t) > \ln(C_t/D\ t)$

As CT_t is supposed to be proportional to the population number the ratio C_t/CT_t is an estimate of the probability that a fish of age t will be on the fishing ground and be retained if it encounters the gear, i.e. C_t/CT_t can be used as an estimate of the resultant ogive S_t.

CT can be predicted by above equation, modified into:

$CT_t = \Delta\ t*\exp(a - Z*t)$

Thus, the ogive can be estimated by:

$$St = \frac{C_t}{CT_t} = \frac{C_t}{\Delta t * \exp(a - Z * t)}$$

The fractions retained of the observed selection ogive are presented in column F of table. In order to obtain the theoretical (estimated) selection ogive the expression for S_t is used in linear form:

$$\ln(1/S_t - 1) = T1 - T2*t$$

Above equation enables us to estimate the parameters T1 and T2 by linear regression. Columns B (x) and G (y) of table contain the inputs for this regression. (Columns C, Δ and E contain the results of the catch curve analysis used to calculate the dependent variable, y, column G.) Column H contains the estimated selection ogive. Only values of S_t less than 1 (column F) can be used in the expression ln(1/S-1) (column G). Carrying out the regression analysis we find:

$$a = T1 = 4.396 \text{ and } -b = T2 = 3.701$$

which gives, using previous equations:

$$t50\% = T1/T2 = 1.1877 \text{ years}$$

$$t75\% = (T1 + \ln 3)/T2 = 1.4846 \text{ years}$$

$$L50\% = 50*(1 - \exp(-0.3*(1.1877 - 0))) = 15.0 \text{ cm}$$

$$L75\% = 50*(1 - \exp(-0.3*(1.4846 - 0))) = 18.0 \text{ cm}$$

The example of table is a hypothetical one constructed to give the results of Pigs. Because the data are ideal hypothetical data there is a perfect agreement between the observed fractions retained (column F of table) and the theoretical fractions retained (column H).

The exercise provides a check on the adequacy of the choice of points used in the regression analysis for the estimation of Z. The conclusion to be drawn from Table is that the first length group to be used in the estimation of Z should have been 27-29 cm, as this group is the first one under full exploitation.

However, because the logistic curve never attains the value 1 the concept of "full exploitation" is determined by the number of decimals in the table.

Taking into account that the logistic curve is an approximation to the real selection curve one cannot expect to get a precise estimate for the first length under full exploitation.

If we are somewhere in the "near neighbourhood" of 1 the choice of first length group in the catch curve regression is likely to be good enough.

As emphasized in the introduction to this section, the results of the method described should be treated with a certain reservation.

Gear Selectivity and VPA Methods

Gear Selectivity and Fishing Mortality

Fishing mortality, F, is clearly related to the selection ogive. When $S_L = 0$ fishing mortality must be zero and when $S_L = 1$ fishing mortality is at its highest level. The obvious relationship between fishing mortality and selection is:

$$F_L = Fm*S_L$$

where Fm is the "*maximum fishing mortality*". Thus, F, as a function of length has the same shape as S, but it has a different level.

We consider F a continuous function of length, L. In practice, however, it is often convenient to replace the continuous function by a step function, where F is assumed to remain constant within each length group.

The continuous selection curve S_L may also be approximated by a step function, S(j), in which the value for length group no. j is S((L1 +L2)/2), where L1 and L2 are the lower and upper limits of length group no. j. When we use the length group index, j, as argument rather than the length L, we may write a step-function model for total mortality, Z;

$$Z(j) = M + Fm*S(j)$$

where M is the natural mortality coefficient (here assumed to remain constant for all length groups), S(j) is the step function of the selection ogive and Fm the maximum fishing mortality. If Z, M and Fm are known the selection can be estimated by (Pope *et al.*, 1975 and Hoydal *et al.*, 1982):

$$S(j) = F(j)/Fm$$

where

$$F(j) = Z(j) - M$$

Figure shows F(j), Z(j) and S(j) as functions of length. When we work with a step function rather than with a continuous logistic curve the selection is given as an array of S-values and this array can replace the mathematical expression or the array may be applied to estimate the parameters of the logistic curve.

Actually, an array of S-values is a more versatile way of presenting selection ogives as no assumptions have to be made about the underlying mathematical expression.

Estimation of Selection Curves from Cohort Analysis

The various kinds of cohort analyses produce an array of estimates of F-values (the so-called "*fishing pattern*"), by age group or by length group.

These F-values supply data for a gear selectivity/recruitment curve which is obtained by:

$$S(i) = \frac{F(i)}{MAX\{F(i)\}}$$

where F(i) is the fishing mortality for age or size group i and MAX{F(j)} is the maximum value of the fishing mortality among all age or size groups applies to any gear or any combination of gears combined with any recruitment curve (Hoydal *et al.*, 1980 and 1982).

The method makes no assumptions on the type of gear or on how the fish are caught.

Thus, the selection curve of a gear can be estimated from catch data only. The actual results of fishing operations and is therefore called the "*effective mesh size*", i.e. the observed selection/recruitment parameters. The concept of "effective mesh size" also applies to gear without meshes, such as hook and lines.

This approach has a number of advantages over methods where selection curves are derived from measurements of gear characteristics, e.g. mesh size. Let us, for example, consider the trawl. If we assume (as is often done) that the gear selection curve is determined by the mesh size in the codend only, then two fishing boats using gears with the same mesh size should have the same gear selection ogive.

However, this is likely to be the case only if the two boats operate the gear in exactly the same way.

For example, if one boat takes hauls of 5 hours duration and the other boat only uses one hour for a haul the selective properties may be different because of clogging of the net by the catch. Also the towing speed may influence the selectivity. Higher speed may make the meshes more elongated and cause a lower selection factor.

Sampling

The ideal basis for fish stock assessment is data that fully represent a stock, at least from the moment that it has recruited to the fishery, without any systematic errors or biases. Although it may not be possible in practice to obtain data of such quality, it should be the aim of any programme for the collection of data on a fishery, to obtain samples that fully represent the population under investigation, to know the sources of bias and to find means to correct for those biases. The full theory of sampling is dealt with in many textbooks, which are widely available, such as Raj (1968), Som (1973) or Cochran (1977). In the first part of this Chapter, some basic aspects of sampling will be discussed, with emphasis on random sampling. The second part deals with an example of a typical tropical demersal fishery, where many species are landed, partly sorted, for human consumption and partly mixed in the form of by-catch (for fish meal production or other purposes).

The aim of this example is to show how complex a sampling scheme may have to be in order to get a representative sample of a particular fishery and how raising factors should be applied to reflect the total fishery for a particular species. This example was created on the basis of experience in FAO/DANIDA follow-up courses on fish stock assessment, where many data sets on tropical fisheries turned out to be incomplete or biased, due to errors in the design and/ or execution of sampling schemes.

Good sampling requires large and long-term investments in terms of manpower and general expenses. Therefore, it is important that sampling programmes are designed in such a way that they provide the data needed for the assessment and management of important species and fisheries, and that these data meet the standards as formulated by international and national working groups. The sampling principles dealt with in this Chapter also apply to the procedures for sampling catches on board of research vessels.

7

Conditioning the Parent Fish

Before placing the parent fish together for spawning, they should be conditioned. Conditioning is feeding the fish a variety of healthy foods to get them in top condition for spawning. Many species can be conditioned using a well-balanced flake food, though others should be conditioned on live foods such as brine shrimp, insect larvae, and flying insects. The parent fish can be while conditioning. This way, when the fish are reintroduced, they are eager to spawn.

Spawning Tank

Though some species readily spawn in the aquarium, the eggs or fry often do not survive because of predatory parents or other fish. Often the fry die because of unfavourable, unclean water conditions. Many species that practice brood care will harm other tank mates in attempting to guard the eggs. Because of all these problems; most aquariasts who breed fish use a separate spawning tank. The spawning tank should be like the hospital tank with protected heater so the the fish are not burned; a slow-moving filter (sponge filter), so the eggs or fry are not sucked up; and good aeration. Depending on the spawning method, the spawning tank can be set up in a number of different ways.

Egg-scatterers: Because egg scatterers often eat their own eggs, the spawning tank has to be set-up so the eggs fall out of the reach of hungry parents. A 5-10 gallon tank is sufficient for spawning for most eggs scatterers.

For egg scatterers like barbs and danios, which lay non-adhesive eggs, the spawning tank can be furnished with a substrate consisting of two layers of marbles or a nylon netting just above the tank floor. As the eggs are laid, they fall through the marbles or the netting out of the reach of the parents. After spawning is over, the eggs or the parents can be removed.

For egg scatterers that lay adhesive eggs like tetras, the spawning tank should be furnished with a substrate. The tank should be planted with fine-leafed plants. The eggs are laid amongst plants, and adhere to the fine-leaves. The parents should be removed after spawning. Rainbowfish also lay adhesive eggs, however, most species spawn continuously over a period of several weeks. Thus, the eggs or the plants that the eggs are attached to, should be removed daily and placed in the rearing tank. Larger rainbowfish require a spawning tank of 20-30 gallons.

Egg-depositors: Depending on the type of egg depositor, the tank should be furnished differently. For those egg-depositors that care for their young, the parents can remain in the tank after spawning. Substrate spawners, depending on the species, should be given a tank furnished glass panes, broad-leafed plants, or flat stones for spawning sites. Some species such as Discus and Angelfish prefer vertical surfaces. For cavity spawners, flower pots turned on their side, coconut shells, and rocky caves are suitable spawning sites. The tank should be furnished with either live or plastic plants to give the fish a sense of security.

Egg-depositors that do not care for their young, should be given a tank furnished with fine and broad-leafed plants, Java Moss, or artificial spawning mops. After spawning the parents or plants with the eggs should be removed. If the plants containing eggs are removed, new plants should be placed in the tank for future spawnings. Killifish eggs often develop best when they are kept in a shallow dish.

Egg-burriers: A peat-moss substrate is one of the best substrates for egg-burying species. The peat moss can be removed after spawning and placed in a plastic bag to be stored for weeks to months (depending on the species). A new peat moss substrate can be placed in the tank for further spawnings. In order to initiate hatching, the stored peat can be immersed in soft water.

Mouth-brooders: Ovophile mouth-brooders can be bred in the main aquarium because the eggs are protected in the mouth cavity. However, it is better to separate mouth-brooders with eggs because of their potentially aggressive behaviour. There are no special breeding tank requirements other than the usual tank set-up for the species.

Larvophile mouth-brooders should be placed in a breeding tank because the eggs are not protected in the mouth, but laid on a surface where the are open to predators.

Nest-builders: Nest-builders should be provided with material with which to build their nests. For bubble-nest builders, fine leafed and floating plants should be provided, and the tank should have no water current to disturb the nest. Species that build nests in the substrate should be given fine gravel or sand.

Livebearers: Small livebearers can be bred in breeding traps where the newborns fall out of the reach of the mother.

However, a more preferable set-up is a separate, heavily planted tank. As the female releases the young, she can be fed so that she is not to eat the fry. As soon as all the young are born, remove the mother.

Stimulating Spawning

One of the best ways to induce fish to spawn, especially difficult-to-spawn species, is to simulate natural conditions. Among factors that encourage fish to spawn are the environment, the food, and the rainy season.

Water Conditions: The right water conditions are among the most basic requirements in spawning fish. Thus the water conditions should be similar to those in the natural environment of the species. By following the suggestions under "breeding" or "water" in the species descriptions, approximate natural water conditions can be found. Another important environmental conditions is the the right tank set-up including hiding places, spawning sites, lighting, water current, and social conditions (schools).

Food: The right foods are important to encouraging spawning. Without proper foods, natural conditions cannot be entirely recreated. Some of the live foods that often can make a difference in spawning success are mosquito larvae and fruit flies.

Stimulating the Rainy Season: Many fish species spawn during the rainy season in nature. By simulating the rainy season in aquaria, difficult-to-spawn species can be induced to spawn. Rains affect the water chemistry, the water height, and the water temperature.

In order to simulate rainy conditions, the tank level should be reduced to half its normal height. Each day add 5% of the tank volume. The water added should be very soft and slightly cooler than the tank temperature. To simulate the rain, use a drip system or spray bar for several hours a day and simulate storms by flashing light in a dark room. Thunder can be created by playing a recording of a storm, a recording of music with bass, or a recording of rattling thin metal cake pans.

The process can be repeated until the fish show signs of spawning preparation. In addition to the "storms," the fish should be fed heavily with mosquito larvae, shrimp, and flying insects. This process will stimulate some hard-to-spawn species, but not all. Many species will respond to just one altered water condition, such as a series of water changes with soft water replacement, or a decrease in the water temperature.

Target Fish: "Target fish" can be used to help strengthen the bond between a fish pair. Target fish can be another of the same species or a similar species that is placed in the tank with the breeding pair. This third fish will serve as an object of the aggression of the pair. The pair will work together to chase off the target fish and not fight between themselves. Only use the target fish method in a large tank with plenty of hiding places, so that the target fish is not harmed.

The Rearing Tank

A rearing tank is not required with species that take care of their young, although they are still recommended. For species that do not take care of their young, the rearing tank can be the same tank as the spawning tank as long as the parents are removed.

The rearing tank should have a protected heater, a sponge filter or a filter with a sponge on the intake, and plastic or live plants. In addition, three of the tank sides should be covered with black paper, because a light can encourage fungal infections and cause discomfort for the fry. The water in the rearing tank should be similar to the water used for spawning.

Raising the Fry

The fry develop in a number of minutes (livebearers) to a number of weeks. When the eggs hatch, the larvae that emerge look nothing like the parent fish. Instead, the larvae have a large, yellow yolk sac and are barely able to move, let alone swim. The larvae will feed off the egg sac until all the yolk is gone. Once the yolk sac is gone, the hungry fry will begin to look for food. The fry of small fish can be first fed infusoria, "green water," or egg yolk. Later these fry can be fed larger foods like whiteworms, Daphnia, Artemia nauplii, and ground flakes. These foods are good as a first food for slightly larger fry such as those of cichlids. Once the fish grow larger, larger foods like brine shrimp, larger Daphnia, flakes, insect larvae, and chopped Tubifex worms are accepted.

25-50% of the water in the rearing tank should be changed daily. Be sure that the "new" water added has characteristics like the water taken out, because fry are sensitive to sudden changes in the water chemistry. The fry should be fed several times a day. Many species need periodic sorting by size, so that larger fish do not cannibalize smaller fish. With favourable water conditions, regular water changes, and generous feeding, the fry should grow quickly. Cull unhealthy and deformed fish.

Fishkeeping

Figure: *An aquascaped freshwater aquarium*

Fishkeeping is a popular hobby concerned with keeping fish in a home aquarium or garden pond. There is also a fishkeeping industry, as a branch of agriculture.

Types of Fishkeeping Systems

Fishkeepers are often known as "aquarists", since many of them are not solely interested in keeping fish. The hobby can be broadly divided into three specific disciplines according to the type of water the fish tolerate: freshwater, brackish, and marine (also called saltwater) fishkeeping.

Freshwater Aquarium

A freshwater aquarium is a receptacle that holds one or more freshwater aquatic organisms for decorative, pet-keeping, or research purposes. Modern aquariums are most often made from transparent glass or acrylic glass. Typical inhabitants include fish, plants, amphibians, and invertebrates, such as snails and crustaceans.

Figure: *Tropical freshwater aquarium*

Freshwater fish may be either coldwater or tropical species. Although freshwater aquariums can be set up as community tanks, coldwater and tropical fish are generally not mixed due to incompatibilities in temperature requirements. Coldwater aquariums house goldfish and other species that do not require a heating apparatus. Warmer temperatures would actually increase their metabolism and shorten their lifespan. For a tropical fish tank, maintaining a warm environmental temperature ranging between 75 to 80 °F (24 to 27 °C) enables the fish to thrive.

Aquariums may be decorated with sand or gravel, live or plastic plants, driftwood, rocks, and a variety of commercially made plastic sculptures. The smallest aquariums are fish bowls, but these are not recommended for most fish as they are generally too small.

History: The earliest known aquariums were artificial fish ponds constructed by the ancient Sumerians over 4500 years ago. The ancient Assyrians, Egyptians, and Romans also kept fish in ponds for food and entertainment purposes. The ancient Chinese were the first culture to breed fish with any degre of success. They raised carp for food around 2000 B.C., and developed ornamental goldfish by selective breeding. Goldfish were introduced to Europe during the 18th century.

In the later 18th century, widespread public interest in the study of nature was awakening, and fish were kept in glass jars, porcelain containers, wooden tubs, and small artificial ponds. It was during this time that zoologist and botanist, Johann Matthaeus Bechstein, kept a large number of fishes and amphibians and laid down the foundation for aquarium and terrarium science. The concepts of the proper aquarium and terrarium were developed later by Nathaniel Bagshaw Ward in 1829. During the 19th century the idea of the "balanced aquarium" was developed. This approach was an attempt to mimic a balanced ecosystem in nature.

According to this method, fish waste could be consumed by plants, and plants along with the air surface of the water could supply oxygen for the fish. In 1869, the first tropical fish (the Paradise Fish) was imported from Asia. In these days, tropical tanks were kept warm by an open flame. Because early filters were noisy and expensive, fishkeeping was a hobby reserved for wealthy, scientifically inclined individuals.

Figure: *A heavily stocked fancy goldfish aquarium*

In 1878, Rear-Admiral Daniel Ammon brought the first tropical fish from the Far East to the United States which lead to a decline in the popularity of goldfish. In the early 20th century, aeration, as well as particulate and charcoal filtration was introduced. The undergravel filter was introduced in the 1950s. By this time, the old idea of the balanced aquarium was viewed as unattainable and unnecessary by many people in the aquarium hobby, but it made a comeback at the end of the 20th Century with the rising popularity of the planted tank.

Today fishkeeping has become a popular hobby that almost anyone can do. Aquarium fish are both wild-caught and bred in Asia and Florida. Captive-bred species are inexpensive and widely available, and are less likely to be infected with diseases or parasites. Unfortunately, successive generations of inbred fish frequently have less colour and sport smaller fins than their wild counterparts.

Fundamentals

A typical household freshwater aquarium set-up, apart from its aquatic tenants, consists of furnishings such as a gravel substrate, live or plastic plants, rocks, driftwood, a backcloth or background, and other decorations. Other equipment includes a canopy or hood as an aquarium cover, an aquarium stand or base, lighting accessories, a heater, a thermometre, air pumps, filtration apparatus, airstones, fish food, a fish net, water conditioner, water quality testing kits, a siphon hose or gravel cleaner, and a bucket for water changes.

Surface area and height are important in the set-up and maintenance of a living biotope. The surface area contributes to providing superior in-tank oxygenation and it also facilitates the creation of attractive aquatic themes. Freshwater environments benefit more from short and wide aquariums, due to the larger surface area they present to the air; this allows more oxygen to dissolve in the water, and the more oxygen there is, the more fish you can keep.

In general, a larger-sized aquarium provides a more stable water-world and the hobbyist can also acquire a greater number of fish. A large aquarium can also enhance aesthetic value. With regards to material, an all-glass aquarium is preferable due to its reasonable cost and its superior ability to resist scratches and discolouration. Indoor aquariums are normally placed far from windows, heating and cooling ducts of the house becaus direct sunlight and temperature changes can negatively affect the aquatic environment. Overexposure to sunlight leads to rapid algae growth inside and outside the tank. Sudden temperature variations are harmful to fish.

Themes

Fish come in a large variety of species, from several different geographical regions. Most aquarium fish originated in Central

America, South America, Africa, or Asia. Fish can be kept in different combinations of species and in different kinds of aquatic environments. Four common themes include the community aquarium, the goldfish aquarium, the African cichlid aquarium, and the planted aquarium. A *community aquarium* refers to the mixing of fish and plants from different geographical areas with an emphasis on the colour and hardiness of the specimens. An example would be the combination of gouramis, tetras, and rasboras with a selection of hardy plants such as Hygrophila difformis, Hygrophila polysperma, and Vallisneria spiralis. Choosing fish that are peaceful and compatible with each other is important in a community tank.

A *goldfish aquarium* can be set up as a unfurnished and bare-bottom tank to emphasize the bright colouration of the fish. A combination of different varieties of goldfish and decorations that contrast with the vivid colours of the fish would make an attractive display. Live plants are not usually grown with goldfish, except for hardy, oxygenating plants like Egeria, because goldfish regularly disturb the substrate. They may also feed on softer-leaved plants. Plastic plants can be used instead.

An *African cichlid aquarium* commonly consists of Lake Tanganyika or Lake Malawi cichlid varieties, and generally requires a large number of rocks combined with a substrate of fine gravel or sand. The rocky environment should provide numerous caves and hiding places. Because cichlids, like goldfish, disturb the substrate by digging, plastic plants should be used as a substitute for live plants. However, real plants like Vallisneria or Anubias can be tried in a cichlid tank.

A *planted aquarium* emphasizes living plants as much as, or even more than fish. Large groupings of plant species such as Hygrophila, Limnophila, Rotala, Vallisneria, Echinodorus, and Cryptocorynes with a limited number of fish is a good example of a planted tank. It is important to select fish that will not damage the plants, such as small tetras, dwarf gouramis, cherry barbs, zebra danios, and White Clouds. Planted tanks may include CO_2 injection and a substrate fortified with laterite or, in the case of a low tech aquarium, a layer of potting soil under the gravel to provide nutrients for the plants.

A *biotope aquarium* is an aquarium that is designed to simulate a natural habitat, with the fish, plants, and furnishings all representative of a particular place in nature. Because only species that are found together in nature are allowed in a true biotope aquarium, these tanks are more challenging and less common than the other themes.

Marine Aquarium

A marine aquarium is an aquarium that keeps marine plants and animals in a contained environment. Marine aquaria are further subdivided by hobbyists into fish only (FO), fish only with live rock (FOWLR), and reef aquaria. Fish only tanks often showçase large or aggressive marine fish species and generally rely on mechanical and chemical filtration. FOWLR and reef tanks use live rock, a material composed of coral skeletons harbouring beneficial nitrogen waste metabolising bacteria, as a means of more natural biological filtration.

Marine fishkeeping is different from its freshwater counterpart because of the fundamental differences in the constitution of saltwater and the resulting differences in the adaptation of its inhabitants. A stable marine aquarium requires more equipment than freshwater systems, and generally requires more stringent water quality monitoring. The inhabitants of a marine aquarium are often difficult to acquire and are usually more expensive than freshwater aquarium inhabitants. However, the inhabitants of saltwater aquariums are usually much more spectacular than freshwater aquarium fish.

Marine Fishkeeping History: The first saltwater tanks were Venetian glass jars where the Romans kept anemones outdoors, but these systems were very short lived. The first personal saltwater fishkeeping began on a wider scale in the 1950s, starting with the basic rectangular glass aquariums (usually 20 gallon), still popular today. Bleached coral along with a substrate of coarse crushed coral were the norm. Algae, including beneficial types such as coralline algae, were viewed negatively and were generally removed. The clean, sterile tank was viewed as the healthiest. The Aztec Empire had 10 ponds of saltwater aquariums at Texcoco. During the early days of marine aquaria, saltwater was collected at local beaches. Natural saltwater contains many unwanted organisms and pollutants. Aquarium literature of the time suggests that the most commonly kept marine fish were the percula clownfish, sergeant major damselfish, small, brackish-water pufferfish and scats, jewelled blennies, and blue damsels.

Aquariums were equipped with large air compressors, and were heavily aerated and filtered (primarily with undergravel filters, a norm for some time). An ever-growing number of hobbyists experiencing the inconvenience of gathering natural sea water and the concurrent development of analytical chemistry techniques led to research into the chemical composition of sea water. Synthetic salt mixes were developed to replicate the chemical environment of the tropical ocean, including trace elements and salts. This advance made marine fishkeeping popular in areas without access to clean sea water.

Air driven, counter-current protein skimmers and reliable submersible electric heaters were invented in Germany. Various advances in filtration included trickle and hang-on filters, both allowing a more natural equilibrium in the aquarium environment.

The advancement of fluorescent lighting technologies to provide higher output, along with metal halide lighting, enabled the first reef tanks, making it possible to keep corals and other invertebrates without natural sunlight.

More efficient chemical testing allowed aquarists to understand the chemical properties of aquariums. By the 1980s, a biologically-based understanding of how to maintain an artificial ocean environment brought more successful and widespread marine fishkeeping.

Marine Aquarium Components

Figure: *Nano reef aquarium maintained at home*

The major components are an aquarium, usually made from glass or acrylic, filtration equipment, lighting, and an aquarium heater. Marine aquariums can range in volume from less than 80 litres, (<

20 US gal) to over 1,200 litres (300 US gal). Small volumes are more difficult to maintain due to the more rapid changes in water chemistry. The majority of saltwater aquariums are between 160 and 400 litres (40 and 100 US gal).

Types of Marine Aquariums

Marine aquarists typically divide saltwater aquariums into those housing fish only, those housing fish with live rock, and those primarily designed to house corals and other invertebrates (also known as reef aquariums). Many fish hobbyists also divide the types of saltwater tanks based on the water temperatures at which they are kept.

Tropical Marine

The most common type of saltwater fish tank, the tropical marine tank, houses marine animals from tropical climates. Usually kept between 24 to 28 °C (75 to 82 °F), these tanks include tropical reef tanks, as well as fish-only tanks. These tanks tend to have a low concentrations of microscopic plankton and other foods eaten by filter feeders. Most livestock for these aquariums are acquired through commercial means.

Temperate Marine (Coldwater Marine)

Figure: *A temperate marine aquarium*

One of the more obscure types of fish tanks, the coldwater marine tank, holds fish of temperate climates, with temperatures ranging around 10 to 24 °C (50 to 75 °F). While most of these tanks are not as aesthetically pleasing astheir tropical counterparts, colourful species such as the ornate cowfish, blennies and sea anemones can be found. However most coldwater fish are gray or dull in colour.

Most colourful species are found in the western pacific and west coast of North America. These tanks also tend to require extra skill to maintain. A significant diversity of species exists. Since coldwater coral reefs only occur at great depths, most hobbyists are largely confined to fish, sea anemones crustaceans echinoderms, mollusks and Feather duster worms.

A few corals can be found at low depths. Since there are very few commercially available coldwater fish and invertebrates, hobbyists usually have to physically acquire specimens.

The most common way of doing this is by trolling or seining, learned hobbyists use the movement of tides and searching methods to find certain species. Unlike commercially available tropical fish, whose behaviour patterns and tank compatibilities are well documented, coldwater fish require much local ichthyology knowledge in order to maintain them. Many temperate fish have specific local diet requirements, while others, will eat just about any crustacean or frozen foods.

Some fish should not be kept with fish small enough to fit into its mouth, crabs or mollusks. Similarly some crabs can not be kept with some mollusks, while other fish, crabs, mollusks an echinoderms may be compatible with each-other. It takes experience before one can successfully gage the compatibility of the fish and invertebrates in one's area. Due to it being such a localised hobby, not many people go the route of local tanks. However, they are more cost-effective than reef tanks.

Live Rock

Live rock is rock from the ocean that has been introduced into a saltwater aquarium. Along with live sand, it confers to the closed marine system multiple benefits desired by the saltwater aquarium hobbyist. The name sometimes leads to misunderstandings, as the "live rock" itself is not actually alive, but is made simply from the aragonite skeletons of long dead corals, or other calcareous organisms, which in the ocean form the majority of coral reefs. When taken from the ocean it is usually encrusted with coralline algae and inhabited by a multitude of marine organisms.

The many forms of micro and macroscopic marine life that live on and inside of the rock, which acts as an ideal habitat, give it the name "live rock".

Figure: *Bleached coral skeletons, which can be inhabited by micro- and macro-organisms to form live rock*

Origin

Live rock is harvested for use in the aquarium trade from collections in the wild near reefs, where parts may become detached from the main body of coral by storms. It may also be "seeded" from small coralline rocks by an aquaculturalist in warm ocean water, to be harvested later. Live rock can also be seeded by adding base rock to an active reef aquarium that already has live rock. Live rock harbours a wide variety of corals, algae, sponges, and other invertebrates, when they are collected. Corals added to the aquarium later will often become attached to the rock.

Purpose

For the aquarium trade this rock is highly valued not only for the diversity of life it can bring to the closed marine environment, but its function as a superior biological filter that hosts both aerobic and anaerobic nitrifying bacteria required for the nitrogen cycle that processes waste. Live rock becomes the main biological nitrification base or biological filter of a saltwater aquarium. Additionally, live rocks have a stabilising effect on the water chemistry, in particular on helping to maintain constant pH by release of calcium carbonate. Lastly, live rock, especially when encrusted with multiple colours of coralline algae, becomes a major decorative element of the aquarium and provides shelter for the inhabitants. It is often used to build caves, arches, overhangs, or other structures in the tank, a practice known as aquascaping.

In J. Charles Delbeek's article *Your First Reef Aquarium*, he states,

> *"The use of live rock immediately introduces into the aquarium numerous algae, bacteria and small invertebrates all of which contribute to the overall quality of the aquarium water. Live rock has just as much, if not more, surface area for bacteria than a trickle filter. Since live rock in the aquarium contains various types of bacteria, algae and corals, waste roducts such as ammonia, nitrate and phosphate can hve a number of fates. Ammonia, nitrate and phosphate are readily assimilated by algae and photosynthetic corals growing on and in the rock. Ammonia can also be quickly converted into nitrate by the bacteria on and in the rock. This nitrate can be either absorbed by the algae and corals, or it can be denitrified by bacteria in close proximity to the nitrate-producing bacteria."*

Types

There are many different types of live rock. Each is named after the area from which it originated. A large amount of live rock comes from the Southern Pacific region, in areas such as Fiji, Tonga, and the Marshall Islands, as well as from the Caribbean. Each has its own distinct qualities that make it preferable to certain reef aquarists. For instance, live rock from the Fiji region is often porous and large, and rock from the Tonga region is often dense and elongated.

Base Rock

Base rock is a generic term for aragonite rock that has no organisms growing in or on the rock. Base rock is often used as filler rock in the aquarium as it is much cheaper to purchase than live rock. In time, base rock will become colonised by living organisms.

Recently base rock that is mined from inland ancient reefs has become a popular way to keep the aquarium trade going sustainably. This rock is either maricultured and sold as live rock, or can be purchased and grown in the home aquarium.

Base rock can also be made from artificial rock called aragocrete, which is a hand made concrete from combining crushed aragonite, sand, and Portland cement.

After allowing the cement to dry, the pieces are sometimes acid washed to counteract the high pH of the materials, and then allowed to soak in clean water for one or more months. They generally tend to be heavier and less attractive when compared to natural base rock.

Collection Ban

As of August 4, 2008 CITES (Convention on International Trade in Endangered Species) banned the collection of live rock from Tonga, the Marshall Islands, and the Cook Islands. This is due to the over-collecting of rock in these areas. The Secretariat has lifted all of the CITES restrictions in the recently banned countries for exporting species. This includes Tonga, Fiji, and Australia. Most notably, Tonga rock and corals and Fiji corals are available once again.

Filter (Aquarium)

Figure: *Air-driven corner filter*

Aquarium filters are critical components of both freshwater and marine aquaria. Aquarium filters remove physical and soluble chemical waste products from aquaria, simplifying maintenance. Furthermore, aquarium filters are necessary to support life as aquaria are relatively small, closed volumes of water compared to the natural environment of most fish. Animals, typically fish, kept in fish tanks produce waste from excrement and respiration. Another source of waste is uneaten food or plants and fish which have died.

These waste products collect in the tanks and contaminate the water. As the degree of contamination rises, the risk to the health of the aquaria increases and removal of the contamination becomes critical. Filtration is a common method used for maintenance of healthy aquaria.

Biological Filtration and the Nitrogen Cycle

Proper management of the nitrogen cycle is a vital element of a successful aquarium. Excretia and other decomposing organic matter produce ammonia which is highly toxic to fish. Bacterial processes oxidise this ammonia into the slightly less toxic nitrites, and these are in turn oxidised to form the much less toxic nitrates. In the natural environment these nitrates are subsequently taken up by plants as fertilizer and this does indeed happen to some extent in an aquarium planted with real plants.

An aquarium is, however, an imperfect microcosm of the natural world. Aquariums are usually much more densely stocked with fish than the natural environment. This increases the amount of ammonia produced in the relatively small volume of the aquarium. The bacteria responsible for breaking down the ammonia colonise the surface of any objects inside the aquarium. In most case, a biological filter is nothing more than a chemically inert porous sponge, which provides a greatly enlarged surface area on which these bacteria can develop. These bacterial colonies take several weeks to form, during which time the aquarium is vulnerable to a condition commonly known as "new tank syndrome" if stocked with fish too quickly. Some systems incorporate bacteria capable of converting nitrates into nitrogen gas.

Accumulation of toxic ammonia from decomposing wastes is the largest cause of fish mortality in new, poorly maintained, or overloaded aquariums. In the artificial environment of the aquarium, the nitrogen cycle effectively ends with the production of nitrates. In order that the nitrate level does not build up to a harmful level regular partial water changes are required to remove the nitrates and introduce new, uncontaminated water.

Mechanical and Chemical Filtration

The process of mechanical filtration removes particulate material from the water column. This particulate matter may include uneaten food, feces or plant or algal debris. Mechanical filtration is typically achieved by passing water through materials which act as a sieve, physically trapping the particulate matter. Removal of solid waste can be as simple as physical hand netting of debris, and/or involve highly complex equipment. All removal of solid wastes involve filtering water through some form of mesh in a process known as mechanical filtration.

The solid wastes are first collected, and then must be physically removed from the aquarium system. Mechanical filtration is ultimately ineffective if the solid wastes are not removed from thefilter, and are allowed to decay and dissolve in the water.

Dissolved wastes are more difficult to remove from the water. Several techniques, collectively known as chemical filtration, are used for the removal of dissolved wastes, the most popular being the use of activated carbon and foam fractionation. To a certain extent, healthy plants extract dissolved chemical wastes from water when they grow, so plants can serve a role in the containment of dissolved wastes.

A final and less common situation requiring filtration involves the desire to sterilise water born pathogens. This sterilisation is accomplished by passing aquarium water through filtration devices which expose the water to high intensity ultraviolet light and/or exposing the water o dissolved ozone gas.

Materials Suitable for Aquarium Filtration

Numerous materials are suitable as aquarium filtration media. These include synthetic wools, known in the aquarium hobby as filter wool, made of polyethylene terephthalate or nylon. Synthetic sponges or foams, various ceramic and sintered glass and silicon products along with igneous gravels are also used as mechanical filter materials. Materials with a greater surface area provide both mechanical and biological filtration. Some filter materials, such as plastic "bioballs", are best used for biological filtration.

With the notable exception of diatom filters, aquarium filters are rarely purely mechanical in action, as bacteria will colonise most filter materials effecting some degree of biological filtratio. Activated carbon and zeolites are also frequently added to aquarium filters. These highly porous materials act as adsorbates binding various chemicals to their large external surfaces and also as sites of bacterial colonisation.

The simplest type of aquarium filter consists only of filter wool and activated carbon. The filter wool traps large debris and prticles, and the activated carbon adsorbs smaller impurities. These should be changed regularly at suitable intervals. This is particularly important in the case of activated carbon filters, which may re-release their adsorbed contents in large (and therefore harmful)doses if they are allowed to saturate. Activated carbon adsorbs toxins on the surface of the carbon. It can be reactivated by boiling in clean water, and draiing away the tainted water, and can then be re-used.

Types of Aquarium Filters

***Figure:** A commercially available canister filter*

Numerous types of aquarium filters are commercially availabl, including:

External Filters

External filters remove water from the aquarium which is then pushed (or pulled) through a series of different levels of filter media and returned to the aquarium. They are usually more effective and easier to maintain than internal filters.

Canister Filters

Compared to filters that hang on the back of the aquarium, canister-style external filters offer a greater quantity of filter materials to be used along with a greater degree of flexibility with respect to filter material choice. Water enters the canister filled with the chosen filter material through an intake pipe at the bottom of the canister, passes through the material, and is fed back to the aquarium through the return pipe. Water is forced to circulate through the filter by a pump typically installed at the top of the canister. It is important to note that canister filters are sealed, fully flooded systems, meaning that the aquarium, intake pipe, filter interior and the return pipe form a continuous body of water. In this configuration both the intake and return path form two siphons, which precisely counterbalance each other. Under these circumstances, the filter pump does not have to spend any effort to lift the water back to the aquarium, regardless of how high the latter is installed above the canister. The pump should only be powerful enough to push the water through the filtering material as well as overcome the drag in the intake and return pipes. This makes canister filter pumps virtually insensitive to the height difference between the aquarium and the filter (although exceeding the manufacturer-specified height limit can lead to leaks).

Benefits of this type of filter are that they can provide a high volume of filter material without reducing the internal space in the aquarium, and that they can be disconnected from the tank for cleaning/ maintenance and replaced without disturbing the aquarium interior or occupants. Also, as a filter with external plumbing, it supports in-line installation of other aquarium equipment, such as water heaters and carbon dioxide diffusers. Such equipment can be removed from the tank and installed in-line into the return pipe of the filter. Disadvantages of canister filters include the increased cost and complexity relative to internal filters and difficulties in cleaning the tubes which transfer water to and from the aquarium. There is also the risk of a leak, which naturally is an issue for any filter placed outside of the aquarium.

Canister filters were initially designed to filter drinking water, under low pressure. Canister filters for aquariums use high water pressure, from a properly powered pump, to force water through the dense filter media. A pump can draw water from an under-gravel filter, and run it into a canister for double filtration.

Diatom Filters

Diatom filters are used only for sporadic cleaning of tanks, they are not continuously operated on aquariums. These filters utilise diatomaceous earth to create an extremely fine filter down to 1 μm which removes particulate matter from the water column.

Trickle Filters

Trickle filters, also known as wet/dry filters are another water filtration systems for marine and freshwater aquariums. This filter comes in two configurations, one which is placed on top of the aquarium (more rarely seen) and one which is placed below the aquarium (more common). If the wet/dry filter is placed on top of the aquarium, water is pumped over a number of perforated trays containing filter wool or some other filter material. The water trickles through the trays, keeping the filter wool wet but not completely submerged, allowing aerobic bacteria to grow and aiding biological filtration. The water returns to the aquarium like rain.

Alternatively, the wet/dry filter may be placed below the tank. In this design, water is fed by gravity to the filter below the aquarium. Prefiltered water is delivered to a perforated plate (drip plate). Prefiltering may take place in the aquarium via a foam block or sleeve in the overflow, or weir siphon, or it may be prefiltered by filter wool resting on the perforated plate.

The waste laden water from the aquarium spreads over the drip plate, and rains down through a medium. This may be a filter wool/ plastic grid rolled into a circular shape (DLS or "Double Layer Spiral") or any number of plastic media commonly known as Bio Balls. As the water cascades over the media, CO_2 is given off, oxygen is picked up, and bacteria convert the waste from the tank into less harmful materials. From here the water enters the sump. The sump may contain a number of compartments, each with its own filtration material. Often, heaters and thermostats are placed in the sump.

Fluidized Bed Filter

The fluidized bed filter (FBF) is a biological reactor only. The principle is to direct water through a sand (or similar media) bed from below so that the sand becomes fluidized – behaves like a fluid. This mechanism is seen in liquifaction, quick sand, and industrial processes including municipal sewage treatment. The combined surface of all sand particles in the filter is very large, and so there is a large surface for aerobic denitrification bacteria. Therefore the size of the filter can be modest. The filter itself can be internal or external. In its simplest DIY internal version an FBF is very easy to build, with a container, sand, pump, and some plumbing. There are many variables: shape and size of the container, quantity of sand or equivalent, particle sizes, the pump's power, and plumbing.

Figure: *Simple DIY fluidized sand bed filter*

Algae Filters

Algae may be grown purposely, which removes chemicals from the water which need to be removed in order to have healthy fish, invertebrates and corals. This is a natural ("green") filtering method, which lets an aquarium operate the way oceans and lakes operate.

Baffle Filters

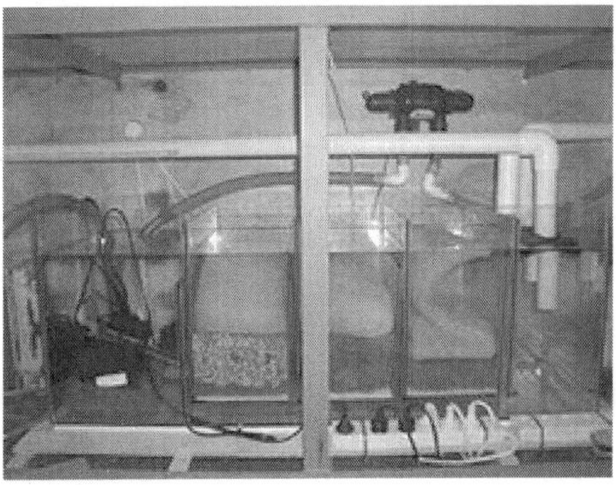

Figure: *A new baffle filter, under a large volume, cichlid aquarium*

Baffle filters are similar to wet and dry, trickle filters in that they are generally situated below the aquarium. This type of filter consists of a series of baffles that the water must pass through in order to reach the pump which is returning water to the aquarium. These baffles then act much like a series of canister filters and can be filled with different filter media for different purposes.

Internal Filters

Internal filters are, by definition, filters within the confines of the aquarium. These include the sponge filter, variations on the corner filter (pictured top right and left), foam cartridge filter and the undergravel filter. An internal filter may have an electric pump and thus be an internal power filter, often attached to the inside of aquaria via suction cups.

Protein Skimmer

A protein skimmer or foam fractionator is a device used mostly in saltwater aquaria to remove organic compounds from the water before they break down into nitrogenous waste. Protein skimming is the only form of aquarium filtration that physically removes organic compounds before they begin to decompose, lightening the load on the biological filter and improving the water's redox potential. Although the process of foam fractionation is commonly known for removal of waste from aquaria, it is, in fact, a rapidly developing chemical process used in the large-scale removal of contaminants from wastewater streams and the enrichment of solutions of biomolecules.

Function

Protein skimming removes certain organic compounds, including proteins and amino acids, by using the polarity of the protein itself. Due to their intrinsic charge, water-borne proteins are either repelled or attracted by the air/water interface and these molecules can be described as hydrophobic (such as fats or oils) or hydrophilic (such as salt, sugar, ammonia, most amino acids, and most inorganic compounds). However, some larger organic molecules can have both hydrophobic and hydrophilic portions.

These molecules are called amphipathic or amphiphilic. Commercial protein skimmers work by generating a large air/water interface, specifically by injecting large numbers of bubbles into the water column. In general, the smaller the bubbles the more effective the protein skimming is because the surface area of small bubbles occupying the same volume is much greater than the same volume of larger bubbles.

Large numbers of small bubbles present an enormous air/water interface for hydrophobic organic molecules and amphipathic organic molecules to collect on the bubble surface (the air/water interface). Water movement hastens diffusion of organic molecules, which effectively brings more organic molecules to the air/water interface and lets the organic molecules accumulate on the surface of the air bubbles. This process continues until the interface is saturated, unless the bubble is removed from the water or it bursts, in which case the accumulated molecules release back into the water column. However, it is important to note that further exposure of a saturated air bubble to organic molecules may continue to result in changes as compounds that bind more strongly may replace those molecules with a weaker binding that have already accumulated on the interface.

Although some aquarists believe that increasing the contact time (or dwell time as it is sometimes called) is always good, it is incorrect to claim that it is always better to increase the contact time between bubbles and the aquarium water. As the bubbles increase near the top of the protein skimmer water column, they become denser and the water begins to drain and create the foam that will carry the organic molecules to the skimmate collection cup or to a separate skimmate waste collector and the organic molecules, and any inorganic molecules that may have become bound to the organic molecules, will be exported from the water system.

In addition to the proteins removed by skimming, there are a number of other organic and inorganic molecules that are typically removed. These include a variety of fats, fatty acids, carbohydrates, metals such as copper and trace elements such as iodine. Particulates and other detritus is also removed, along with phytoplankton and bacteria.

This removal is highly desired by the aquarist and is often emphasized by the placement of the skimmer before other forms of filtration; lessening the burden of the filtration system as a whole. There is at least one published study that provides a detailed list of the export products found in protein skimmer skimmate. Protein skimmers are used to harvest algae and phytoplankton gently enough to maintain viability for culturing or commercial sale as live cultures.

Design

All skimmers have key features in common: water flows through a chamber and is brought into contact with a column of fine bubbles. The bubbles collect proteins and other substances and carry them to the top of the device where the foam, but not the water, collects in a cup. Here the foam condenses to a liquid, which can be easily removed from the system. The material that collects in the cup can range from pale greenish-yellow, watery liquid to a thick black tar.

Consider this summary of optimal protein skimmer design by Randy Holmes-Farley:

For a skimmer to function maximally, the following things must take place:

1. A large amount of air/water interface must be generated.
2. Organic molecules must be allowed to collect at the air/water interface.
3. The bubbles forming this air/water interface must come together to form a foam.
4. The water in the foam must partially drain without the bubbles popping prematurely.
5. The drained foam must be separated from the bulk water and discarded.

Also under considerable recent attention has been the general shape of a skimmer as well. In particular, much attention has been given to the introduction of cone shaped skimmer units. Originally designed by Klaus Jensen in 2004, the concept was founded on the principle that a conical body allows the foam to accumulate more steadily through a gently sloping transition. This reduces the overall turbulence, resulting in more efficient skimming. While research into the specific benefits of the design are still being measured, early reviews of many conical skimmers have been positive overall.

Overall, protein skimmers can be classed in two ways depending on whether they operate by co-current flow or counter-current flow. In a co-current flow system, air is introduced at the bottom of the chamber and is in contact with the water as it rises upwards towards the collection chamber. In a counter-current system, air is forced into the system under pressure and moves against the flow of the water for a while before it rises up towards the collection cup. Because the air bubbles may be in contact with the water for a longer period in a counter-current flow system, protein skimmers of this type are considered by some to be more effective at removing organic wastes.

Co-current Flow Systems

Air Stone: The original method of protein skimming, running pressurised air through a diffuser to produce large quantities of micro bubbles, remains a viable, effective, and economic choice, although newer technologies may require lower maintenance. The air stone is most often an oblong, partially hollowed block of wood, most often of the genus Tilia. The most popular wooden air-stones for skimmers are made from limewood (Tilia europaea or European limewood) although basswood (Tilia americana or American Linden), works as well, may be cheaper and is often mre readily avalable. The wooden blocks are drilled, tapped, fitted with an air fitting, and connected by air tubing to one or more air pumps delivering at least 1 cfm. The wooden air stone is placed at the bottom of a tall column of water. The tank water is pumped into the column, allowed to pass by the rising bubbles, and back into the tank. To get enough contact time with the bubble, these units can be many feet in height.

Air stone protein skimmers may be constructed as a DIY project from pvc pipes and fittings at low cost and with varying degrees of complexity.

While this method has been around for many years, many regard it as inefficient for larger systems or systems with large bio-loads.

Venturi

The premise behind these skimmers is that a venturi pump, or aspirator, can be used to introduce the bubbles into the water stream. The tank water is pumped through the venturi, in which fine bubbles are introduced, then enters the skimmer body. This method was popular due to its compact size and high efficiency but venturi designs are now more likely to be included in other skimmer designs rather than as a simple venturi design.

8

Commercial Broiler Production on an Integrated Fish Farm

The broiler farming has only developed for 20 years in China. Large flock and high density farming method is adopted. Broilers grow very quickly. In the feeding condition is favourable they can be sent to market after 8 weeks, usually not over 12 weeks. The average weight of the chicken is between 1,250–1,600 grams. The feedstuffs of broilers should be compound feed with all the essential nutrients. Special attention must be paid to the temperature, and of the house broiler farming,

Chicken House Construction and Main Equipment Farms From the economical point of view, the open-type house with natural lighting and ventilation for chicken is practiced or fish farms.

Usually, it is a one-storeyed building with inverted "V" shape roof. Its length is about 6–8 meters. Its height from the ground to the caves is about 2.2–2.5 meters even 3 meters in the warmer area. Each room has an area of about 100 square meters, which can accomodate about 800 chickens. The width of the house should depend on the amount of chickens raised. Two workers can handle about 3,000–4,000 chickens at one time. The chicken house should face the sun, the ratio of the windows and the ground area is usually 1:8–1:10. The windows in the south should be bigger and nearer to the ground while the windows on the north side should be smaller and farther to the ground in the northern hemisphere. It's better to have a concrete ground 30 cm higher than the ground outdoor so that it will be easier to keep it dry and sterilised.

- *Thermafication:* Thermoelectric umbrellas are commonly used. It is square in shape, about 1–1.1 meters on each side, about 0.7 meters in height, with 45 degrees inclination upward. Each umbrella has a 300 W. Thermoelectric wires and a thermostat which can keep 250–300 chicklings warm. In the cold winter, stoves with pipes should be installed in the house in order to keep the room warmer.
- *Feeding Trough:* There are 2 types of feeding systems, the manuel and the mechanical chain type.

  The feeding trough for chicken is made of wood plate or iron sheet. The size and height depend upon the growth period of chickens. Usually there are small, medium and large sizes. There is a special rod on the trough to prevent the spilling of the feedstuffs and the continuation from chicken manure. There should be a certain number of troughs to guarantee the chicken to be fed evenly. This kind of troughs can also hold wet feedstuffs. Cylindric feeding trough can also be used.

  Chain-driven feeding system: It is a trough made of iron sheet. At one end of the trough, attached a feedstuff box, with a chain in the trough moving transversely, which can deliver the feedstuffs to the whole length of the trough. It is easy to use and save manpower.
- *Waterer:* For chicklings a big opening jar with an aluminium plate beneath can be used. They should be installed around the thermoumbrella in order that the chicklings can easily get water. As the chicken grows, automatic barrel-like plastic waterer can be used instead of jar. It contains clean water for about 100 chickens to drink per day. Some farms use long trough with running water and the results is also promising.
- *Chicken Cage:* The chick transportation cages are made of calciumplastic corrugated paper, which can hold 100 newly-hatched chickens. For the chickens sent to market, they are usually kept in iron wire cage of 90 × 60 × 35 centimeters in size. Each cage can hold about 15–20 live chickens weighing 1.5 kilograms each.

Standards and Formula of Broiler Chicken Feeds

In order to fully utilise the feedstuffs, to cut down the cost of the feed, to accelerate the growth and build up the health of the broiler chicken, we have to set the standards of the nutrition according to nutritional needs. The following nutritional requirements are adopted:

Table: Nutritional Standards for Broiler Chicken

	0–5 weeks old	*Over 5 weeks old*
Metabolic energy (cal/kg)	2800–3000	3000–3200
Crude Protein(%)	20–22	18–20
Ratio of Protein and energy (g/kilo cal)	72	61
Calcium (%)	0.9	0.9
Phosphate (%)	0.65	0.65
NaCl (%)	0.37	0.37
Amino Acids (g/kilo cal)	2.66	
Cysteine	1.24	1.09
Lysine	3.91	3.44
Tryptophan	0.72	0.63
Arginine	4.38	3.75
Leucine	5.00	4.38
Isoleucine	2.69	2.34
Phenylalanine	4.68	4.06
Tyrosine	2.18	1.88
Threonine	2.50	2.19
Valine	3.13	2.66
Histidine	1.44	1.25
Glycine and /or Serine	3.59	3.13

Table: Nutritional Standards of Vitamins and Minerals for Broiler Chicken

	Chicken of 0–8 weeks old
Vitamin A Active (I.U./kg)	1,500
Vitamin D_3 (Chicken I.U./kg)	200
Vitamin E (mg/kg)	10
Vitamin K_1 (mg/kg)	.53
Vitamin B_1 (Thiamine HCl) (mg/kg)	.8
Vitamin B_2 (Riboflavin) (mg/kg)	.6
Pantothenic Acid (mg/kg)	10
Nicotinic Acid (mg/kg)	27
Vitamine B_6 (mg/kg)	3

Contd...

	Chicken of 0–8 weeks old
Biotin (mg/kg)	0.09
Choline (mg/kg)	1,300
Folic Acid (mg/kg)	0.55
Vitamin B_{12} (mg/kg)	.009
Sodium (Na) (%)	0.15
Potasium (K) (%)	
Manganese (Mn) (mg/kg)	55
Iodine (I) (mg/kg)	0.35
Magnesium (Mg) (mg/kg)	500
Iron (Fe) (mg/kg)	80
Copper (Cu) (mg/kg)	4
Zinc (Zn) (mg/kg)	50
Selenium (Se) (mg/kg)	0.1

For easy preparation of the feed, there are vitamin compounds and trace elements additives for broiler chickens manufactured in our country. Wc can just add them to the feed according to a certain proportion.

There are a number of feedstuff companies in the world, they prepare compound feed for broiler chickens. According to the requirements as well as local feed resources, the results are quite promising. The metabolic energy, crude protein, amino acils, calcium, phosphates, vitamins and mineral contents of the feedstuffs' formula can fulfil the nutritional requirements.

Table: Average Feeding Standards for Broiler Chicken

Age by days	*Amount of feed (g) each chicken per day*	*Age by day*	*Amount of feed (g) each chicken per day*
1–5	10.0	31–35	80.0
6–10	20.0	6–40	90.0
11–15	32.0	41–45	100.0
16–20	44.0	46–50	110.0
21–25	58.0	51–55	115.0
26–30	70.0	56–60	100.0

Table: Formula of the Feedstuffs for broiler Chicken Produces by the Wuxi Feedstuffs Company

	0–5 weeks old ratio	***above 5 weeks ratio***
Corn	38	48
Highland Barley	15	10
Soya Bean Cake	5	5
Bran Cake	6	8
Cotton-seed Cake	5	5
Wheat Bran	6	5
Low Grade Wheat Flour	5	
Fish meal	9	9
Peptone	4	2.5
Chinese Scholartree *Leave Powder*	4.5	5
Bone Powder	1.5	1.5
Calcium Carbonates	0.5	0.6
Ferrous Sulphate	0.2	0.2
Table Salt	0.1	0.1
Trace element additives	0.2	0.1
Multivitamin additives	5g.	3 g.
Metabolic Energy (Cal./kg)	3036	3057
Crude Protein (%)	21.57	20.14
Crude Celluose (%)	3.71	2.76
Calcium (%)	1.08	1.06
Phosphate (%)	0.86	0.65

Feeding and Management of Broiler Chicken

The most important thing in the Management of broiler chicken is the formula of the feed. The nutrient of the feed should be all-round. Second, the daily management of the poultry should be emphasized. Good environmental conditions are essential to the growth of the chicken. Better rewards could be obtained from the feedstuffs, if we follow the above-montioned criteria.

Simultaneously in-and-out System

All chickens should start to feed at the same day and come to the market at the same day. After all the chickens are sold, complete cleaning and sterilisation of the poultry farm should be carried out. Then rest for 7–14 days. This period of fallowness can break any cycle of any infectious disease. A "clean start" will be ready for the next batch of chickens, which can be prevented from the infectious diseases of the previous batch. As the growth speed of broiler chickens are not uniform, it is better to modify the simultaneously in-and-out system. Chicklings are received at the same time, but sold according to their body weight. For those growing slower, we keep them one to two weeks more, and sell them while their body weight reaches the requirements.

Good Preparatory Work

After all the chickens are sold, the poultry house and equipment should be completely sterilised and a layer of clean and dry bedding is laid on the floor. Before chicklings are coming, the temperature of chicken nursery should be kept above 80°F, and the temperature beneath the nursing umbrella should be 90–95°F, lamps should be installed in the umbrella in order to attract chicklings to come under the umbrella to keep warm.

The trough and waterers should be also installed around the umbrella. Protecting board is set up outside of the umbrella, thus keeping chicklings in the warm area of the umbrella. Waters and trough should be filled before the introduction of chicklings. Chicklings should come to the poultry house in the morning. In this way, they can learn to eat and drink in the day time.

The lamps should be ligned on in the first two weeks Fresh food and water should be given and the daily consumption of the feedstuffs recorded. The temperature of the nursing umbrella should be checked in the night to prevent chicklings from crowding together if it is not warm enough. No light is needed all the night after 48 hours. The protective board of the nursing umbrella should extend and the temperature lowers one degree F daily.

The air should be kept fresh, water and feedstuffs should be clean. The bedding of the chicken house should be dry and the environment quiet. As chicklings grow, the protective board can be removed on the 7th day to avoid too many chickens gathering together. The trough and waterers should also be changed to larger size as the chickens grow. The nursing umbrella should be gradually raised up in height. If feed is given all day long, the feed and water should not be interrupted. If artificial feeding is carried out, the feeding time should be fixed. Weaker chickens should be separated into groups for feeding. The bedding on the ground should be changed frequently and exposed to the sun as often as possible.

The chicken house should be kept warm in winter and well ventilated and proper measures should be taken to lower the temperature in summer. Chicklings are fed with special diet during the first 5 weeks and then fed with fattening diet gradually. Prevention of diseases should be emphasized during the whole course of raising. When those chickens reach 10 weeks old, with body weight of 1.5 kg, they can be sold first. It is better to catch the chickens in the early morning and each cage can hold 15 chickens in the summer and 20 chickens in the winter. A layer of dry grass is laid on the base of the cage. Tender can during the catching can avoid injuries.

Prevention and Treatment of Common Disease

The raising time of broiler chickens is short, but the density is high. The prevention of disease should stress nursing management, cleanness and hygiene. The prevention program should be outlined ahead of raising. Medical treatment is carried out only hen the prevention fails.

The common chicken diseases are Pullorum disease, coccidiosis avium, infectious bronchitis of chicken, chronic respiratory diseases, variola avium, Newcastle disease, cholera avium, Marek's disease, ascariasis avium and deficiency disease, e.g. lack of certain nutrients, such as vitamins and minerals.

Some of the chicken diseases can be prevented by inoculation of vaccine or by medication. Some big manufacturers have produced mixed whole feedstuffs including vitamins, trace elements and drugs, thus, it can effectively prevent the occurrence of certain diseases. Some of drugs may produce resistance after long application, such as anti-coccidiosis medication. Therefore, the most helpful way to prevent the outbreak of chicken diseases is good nursing management, cleanness and hygeine as well as isolation and sterilisation.

Example of Preventive Project

Marek's vaccine is given to chicklings within 48 hours after hatching. 0.01% furazolidonum is added into the drinking water for 1–7 dayold chicklings to prevent Pullorum disease.

Nose drops of weak toxic Newcastle disease II vaccine are given to the 7 day-old chickens, and they are inoculated with variola avium vaccine at the same time.

Robenidine of 30 ppm is added into the feed after 10 days old to prevent coccidiosis and discontinued one week before marketing.

The beak should be cut in the middle with a beak cutter and hemostasic is done by cautery at 15 days old to prevent bad pecking habit.

H_{120} vaccine of infectious bronchitis is given with the drinking water to 20 day-old chickens at the dilution of 1:500–1000. The dosage is 5–10 ml per chicken.

Subcutaneous inoculation or Newcastle disease vaccine at 45 days old. If ascarid is found, tetrametrasol is given.

Control of coccidiosis: Coccidiosis is the most serious disease of broiler chickens and usually happens after the chickens are 8 weeks old. It is mainly transmitted through manure, especially in the hot and humid season. If the bedding material is thick, the transmission will be more serious because the wet bedding increases the chance of hatching of occyst.

The most effective approach for preventing coccidiosis is to raise chicklings in cages above the ground. Thus, the chickens are isolated from their own excrement. If thick grass bedding is used, the most common approach is some anti-coccidiosic medication given with the feedstuffs right after initial feeding, and should be carried out through the whole period of growth. For instance, 125 ppm of sulfadimethoxine (SDM) or 30 ppm of chlorophenyl-quanide can be added into the feedstuffs. Dinitolmide (Zoalene) can also be added into the feed, up to the concentration of 0.0125%. Attention should be paid to that continuous application of anti-coccidiosis drugs for a period of time may bring about the drug-fast and drug-fast strain. As a result, coccidiosis could break out again then, the treatment of coccidiosis is only one of the approaches in the whole process of prevention. If the result of the first trial is not promising, other medications should be considered. Two kinds of medications can be applied alternatively. Besides the medical control of coccidiosis, cleanness and hygiene of the chicken house and the dryness of the ground bedding are also essential. The addition of vitamin A and K in the feed can increase the resistance of the chicken.

Collection and Utilisation of Chicken Manure

Table: Nutrient Constituents in Chicken Manure (Dry)

	Cage Raising	*Ground Raising with saw dust bedding*	*Ground Raising with dry grass bedding*
Moisture (%)	11.4	12.3	15.5
Crude Protein (%)	26.7	21.0	
Crude Fat(%)	1.76	1.7	2.3
Non-nitrogen Extracts (%)	30.6	30	27.1
Crude Cellulose (%)	13.04	17.2	18.7
Minerals (%)	16.9	16.9	14.1
Calcium (%)	7.8	1.95	2.3
Phosphate (%)	2.2	1.26	0.42

Deodourising of Chicken Manure

The chicken manure is fermented and decomposed a few hours after excretion, and a bad smell will occur. Animals instinctively refuse to eat chicken manure unless they are accustomed to it. It is better to deodourise the chicken manure before feeding.

Adding Ferrous Sulphate: Ferrous sulphate for industrial use ($FeSO_4$ $7H_2O$, green vitriol) has a mild bactericidal function. After mixing it with chicken manure, there will be no fermentation, no decomposition and no foul smell. But ferrous sulphate will absorb moisture, which makes the pulverisation and spreading of the powder difficult. Therefore, the process of defouling is limited. Half amount of chimney dust can be added and mixed, then baked dry at 60–80°C, the mixture will not absorb moisture, and this makes the spreading easier. 7% of ferrous sulphate and 3.5% pulverised coal dust are blended with chicken manure. After drying, it will become odorless chicken manure, which can be tolerated by pig, cow, fish, even chicken itself. Their nutritional value are very high, and can be used up to 20% in the feedstuffs.

Moisture	*Protein*	*Fat*	*Non- Nitrogen Extracts*	*Crude Cellulose*	*Minerals*	*Caloric*
%	%	%	%	%	%	cal/kg
10.35	21.9	1.25	21.33	7.13	30.04	2161

Fermentation Method: If dried grasses are used as bedding, the chicken manure and the dried and pulverised grass (about 14%) are put together into a fermontation pool with dimensions of 2 × 2 × 0.5 m and then, add in equal amount of dried and pulverised grass, silage, wheat bran and a small amount of table salt. The mixture will be filled with water, which occupies 70% and be sincked to 15 cm in height, fermenting for 4 hours in summer. If the temperature is over 40 degrees C, spread out the fermented material. It should be heaped up to 20–30 cm in the spring and fall, let it ferment for about a day, and it should be stacked up to 40–50 cm in height in winter, and covered with a plastic sheet for about 3–4 days. This kind of formented feed has no foul smell at all, but with some wine flavour, good for taste, and pigs like it very much.

Utilisation of Fresh Chicken Manure: The nutritional value of the manure of less than 5-week-old chicken are very high; pigs like it even without any treatment. From the experiments, the manure of the egg-laying hens contains about 60% of the crude protein in the feedstuffs. There are 4 grams of digestible protein in the excreta of the egg-laying hen per day. The fresh chicken manure from the cage-cultured hens can be collected by the manure collector and mixed with the feed immediately for the pigs feed. If pigs are accustomed to such a kind of food, they can be utilised more efficiently. It is better to feed sows, and should not be fed to fatteners, which may influence the quality of pork.

Sterilisation of Chicken Manure: The chicken manure collected contains latent pathogens, sterilisation is needed so as to prevent other animals from chicken diseases. It is considered to be the most appropriate approach in the foreign countries to use methyl alcohol bromide (MeBr) to fumigate manures including bedding. Fumigation can be carried out in the storage basement of silage, or in the special designed tanks.

Another method is to stir up the manure with formalin solution at a dose of 0.5–2% and to put it into use after air-dry.

Chicken manure fed to cow and pig should be limited within 10–30%. If 15–25%, the results will be promising.

Goose Production

Both goose and duck are water fowl. Goose raising is mainly for human consumption. Goose is herbivorous animal, which can utilise some of the green scarse feeds. Not like ducks, geese don't have to use feeds of animal origin. Therefore the cost of the feeds for geese is much lower than that for ducks. The well known Chinese goose has strong ability to graze with early maturation and high output. The ratio of the feeds and meat, that is the reward of feed, is higher for the goose than for the duck. It is more economical to raise goose. Fishpond can also be used to raise goose. The goose droppings go directly into the fishpond, which can fertilize the pond water and can be fed to fish as well. If ducks are raised in the fishpond, they will take small fishes as their food. With geese in the pond is a good natural scenery, "white feathers float on green water; Red palms pluck blue ripples."

Taihu Lake Goose is a Fine Species for an Integrated Fish Farm

Chinese goose is famous for its early maturity and high output. They can be divided into large and small body-types according to their size. Lion-head goose is the largest variety of Chinese goose, and also one of the largest varieties in the world. Matured goose weighs about 10–12 kilograms, gosling grows very quickly, 75–90 day-old meat goose can weigh about 5–7.5 kilograms. But their ability to seek food is poor and they need more feedstuffs, bringing forth only 25–35 eggs per year. Taihu goose is famous for small body-type white goose in China, its size is not big, the matured goose only weighs about 3.5–4.5 kilograms. But they mature earlier and can lay about 80 eggs per year. High yield species can lay more than 100 eggs per year. They have a good ability to seek food and the consumption of foodstuffs is less; therefore, Taihu goose is better than others from the economical point of view. We take small size, early matured, high yield Taihu goose as maternal-line and large size lionhead goose as paternal-line, through artificial insemination, cross matching and heterosis to produce meat gosling.

Grazing of Meat Gosling

Spring is the grazing season for the meat goslings. The weather in spring gradually becomes warm and grasses sprout. It's good time for the growth and grazing of the goslings. After wheat harvest, goslings can graze in the wheat field and they can seek the remaining wheats to fatten themselves, therefore, less feedstuffs are required. Only about one kilogram of feedstuffs is required for each gosling up to 70 days old.

The newly -born goslings are very timid and afraid of cold and like to get together, therefore, they should be kept warmer and divided into small groups. They have to be fed for at least six times per day at a fixed time, and midnight feeding is essential.

Feedstuffs can be fragmentary rices, chopped green cabbages or green grass. After 4 or 5 days, they are allowed to graze and the distance of the grazing can increase gradually. After 15 days they can go out for camping.

They should be kept in the hut for the night to prevent injuries from the wild animals. During the fattening stage, if the weather is hot, they are kept in the rivers, lakes and ponds for the night. Up to 70 days, the average weight reaches 2.5 kilograms, and then they can be marketed.

Management and Feeding of Breed Goose

Goose is a water fowl. Integrated fish farm can use fishpond to raise goose. A simple goose house can be erected on the bank of the fishpond. The goosehouse should face the south, the southern side is open or enclosed by low walls. There should be a dry run of at least 5 meters in width for feeding and resting. A slope of less than 30 degrees is to connect the dry run with the fish pond.

The fishpond is enclosed with hedges about 60 centimeters in height. The ground of the goose-house is 20 centimeters higher than the outside and grass is used as bedding to keep the house dry. In one corner of the house, straws are stacked for laying eggs.

Each square meter of the house can hold about 4 Taihu geese, but the water area should be as large as possible. In order to get high fertilization rate of seed egg, there should be 15 male geese matched with 100 female geese. Taihu goose can utilise green fodders; therefore, during non-egg laying season, raising breed geese depends on grazing with a small amount of blighted rices as supplementary food.

The feedstuffs should be increased upon the egg-laying stage. With the increasing production of the eggs, the ratio of fine feeds should increase in proportion and special attention should be paid to the green feeds and minerals supplies.

In order to promote the egg-laying of the breed geese, lamps should be put on in the night so as to prolong the lighting time from 12 hours to 15 hours. Grazing should be performed as much as possible. The feedstuffs ingradient and amount of feeding of the Taihu goose would be as follows:

Table: The ratio of Feeds for the Breed Goose

Month	*Daily feeding amount*	*feed ingradient (%)*					*Metabolic energy (Cal/kg)*	*crude protein (%)*
		blighted rice	*bran*	*bran cake*	*Frag rice*	*barnyard grass*		
Jun-Aug	150	50	50				557	3.78
Sept-Nov	250	50	20	15	15		842	7.34
Dec	250	40	20	15	15	10	1346	9.75
Jan-Mar	200	40		20	20	20	1746	9.75
Apr-Jun	175	30		20	20	30	1748	10.28

In the above-mentioned ratio of the feedstuffs, the nutrients, especially the crude protein and metabolic energy are too low, but it is due to limitation of the food source and raising cost, this is the best ratio we can obtain. As Taihu geese have good propensity to graze and good ability to utilise coarse feed, they can adjust their intake according to the requirements of the energy and protein. Thus, they can overcome the disadvantages in the formula. If we can improve the formula of feeding, the egg-laying rate can be raised.

Common Goose Diseases

Matured geese are rather strong and have good resistance against diseases and do not easily get sick. But the resistance of the goslings are very weak.

If they are kept in large goosery with dense population, any little slip in the management, that is against the physiological requirements of the goslings will cause death. The best preventive approaches to the goose disease are good rearing management, sanitary environment and proper vaccination

The common infectious diseases among the goosery are gosling pest, cholera avium and yolk peritonitis. Gosling pest vaccine is now available in China to prevent its occurrence. The mother goose should be inoculated with diluted 1:100 gosling pest vaccine one month before egg laying, once a year, and then the goslings hatched will have a strong immunity. Regular inoculation of cholera avium vaccine can prevent cholera of the geese. If the goosery is already infected, streptomycin should be injected at a dose of 100,000 IU for each goose at four-hour interval for three times, and the result is promising. Yolk peritonitis is usually occuring during the egg-laying season and there is no such vaccine for prevention now.

However, streptomycin can be used for treatment, and the whole goosery should be fed with furazolidonum mixed with the feeds at a dose of 25 mg for each goose for 3–4 consecutive days. No matter what communicable disease is, besides medical control, other measures such as sterilisation, isolation and deep bury of carcase ought to be taken timely to prevent the spreading of disease.

Duck Production

Chinese people have a tradition of eating duck eggs and processing them into salted or preserved eggs. Raising of egg-laying duck needs less and simpler equipment. They mature earlier, produce more eggs, and the size of the egg is rather large.

Ducks live in groups, and feed on natural feeds through grazing. Therefore, the raising of egg-laying duck is more profitable than that of egg-laying hen. On an integrated fish farm, grow-out pond can be used to raise egg-laying duck. Duck is an omnivorous water fowl which can use the by-products of the crops and animal feeds as well. Duck manure has higher nitrogen contents, which doubles that of goose manure, and the phosphate contents are nearly twice that of goose manure.

1. Shaoxing duck is one of the good egg-laying species in China: Its somatotype is rather small, the average weight of the matured duck is about 1.25– 1.5 kilograms. Its ability to seek food is rather strong, so it needs less artificial feed. It matures earlier and usually lays eggs at the age of 120 days. 250–300 eggs can be produced every year. High productive groups can give more than 300 cggs/duck per annum. Each egg weights about 55–65 grams. We have introduced drake of Khaki-Compbell strains from Holland and crossed them with Shaoxing female duck. The hybrid vigor is rather obvious: the first generation hybrids start to lay eggs after 102 days, averagely producing 287.7 eggs per year. Each egg weighs about 64.5 grams. It can produce about 18.8 kilograms of eggs per annum. The economic benefit is about 50% higher than that of Shaoxing duck.
2. Maximum utilisation of natural resources for grazing: Duck is an omnivorous water fowl. Their ability to seek food is very great; therefore, they should be raised by grazing as much as

possible. We can utilise the natural resources in the rice fields, lakes and marshes, trenches and ditches or river shoals. Egg-laying ducks usually hatch their ducklings in the early fall, the paddy field can be used for grazing. Ducklings eat grass, snails, small fishes and shrimps and injurious insects in the paddy field. After the rice grows up, transfer the ducks into the rivers, channels, ditches, lakes or marshes for grazing. After the harvest of rice crops, duck can be driven back to rice paddy fields where there must be some left-over rice, barnyard grass, snails and mole crickets. The natural feeds are seasonally available, duck raising must take advantage of natural resources so as to save artificial feeds. When grazing stops by November, ducks start to lay eggs.

3. Raising Ducks in Fishponds: Fishponds on an integrated fish farm can be used to raise ducks, but ducks eat small fishes, and they dive into water competing with black carp for snails. Therefore, we had better raise the ducks in grow-out ponds with much bigger water area. The construction of duck coops is similar to the construction of goose house, but egg-laying ducks are more sensitive to cold in comparison with geese. Lower temperature will affect the egg laying, so duck coops should be kept warm, there should be windows and doors on the southern wall to keep coops cool in summer and warm in winter.

If there are some lakes, marshes, river shoals with plenty of natural feeds nearby the duck farm, grazing should be utilised to its utmost. Otherwise, they could be bred in suck shed, moving around in dryrun, wetrun and fishpond for food and rest, and all the egg-laying ducks will live mainly on artificial feeds.

Each Shaoxing duck has to be fed with 110–120 grams of mixed feedstuffs, including blighted rices, barnyard grass, rice bran and fragmentary rices, and 50 grams of animal feeds, such as fresh fishes and pupae. Besides 100–200 grams of green vegetables for each duck per day, snails and Corbicula spp. are often to be supplemented to meet their calcium needs.

The green fodder should be chopped and fresh fishes should be cooked and then be mixed with green fodder. Feeds are given three times daily at set time. Shaoxing duck is rather nervous, easily to be disturbed, which may influence their egg-laying ability. Therefore, we should keep the environment very peaceful and quiet. Although egg-laying duck is a water fowl, they have to rest on land and thus, the duck house should be dry and grass bedding should be thick in winter in order to keep it warm and dry.

Prevention and Treatment of Common Diseases

The most important infectious diseases of egg-laying duck are duck pest and fowl cholera. We now have vaccine for preventive inoculation against the pest.

The results are very promising, while the immunological effects of the fowl cholera is still unsatisfactory. To treat fowl cholera, each egg-laying duck is injected intramuscularly with 50,000 IU of streptomycin at an interval of 4 hours for three consecutive times, then followed by oral ingestion of 0.2 grams of sulfadimethoxinum daily for 3 consecutive days. Sterilisation, isolation and deep bury of carcase should be taken to prevent the spreading of the disease.

Pig Farming

Pig is an omnivorous animal, the length of pig's digestive tracts is 14 times of its body length. Feedstuffs are much better utilised by pig than by chicken. Chinese pigs can tolerate and fully utilise coarse fodder, even chicken excrement including the bedding. After being accustomed to it, pigs can take a large amount of these fodders. Pig has become one of the main animals in the complete set of "fish-livestock-poultry" system on an integrated fish farm.

Pigsty on the Pond Dyke

Pigsties usually are built on pond dyke on an integrated fish farm, so that the excreta of pigs can be directly flushed into the pond. Because the pond dyke is not too wide, the pigsty is usually of single row, facing the south with inverted "V" -shaped roof.

The depth of the pigsty from south to north is 4 meters, the width of every house 3 meters. There is a veranda about one meter in width on the southern side. Each house has an area of about 11 square meters and can hold 10 fattening hogs. The height from the ground to the eaves can be 2–2.2 meters. In order to reduce the expenses, no wall is needed on the southern side, but only a fence about 1.2 meters in height. The cement ground is a little higher in the northern side, and slightly sloping down toward the south.

Thus, pigs' excreta can easily pass out directly into the fishpond through discharge ditches. This kind semi-open type pigsty has the advantage of less building cost, good ventilation, sufficient sun-light, cooler in summer and warmer in winter.

The feeding trough is installed on the southern side and pigs sleep on the northern side. Thus, the ground can be kept dry and it is convenient for operation and management.

Utilisation of Hybrid Vigor in Pig Farming

Taihu pig bred in the Wuxi district can copulate when it reaches 30 kilograms in weight at 5 months old. Each litter has about 15.5 piglets on average.

Matured sow weighs about 125 kilograms. They mature earlier and give more piglets for each litter.

The quality of the pork is good, less fodders are used and they can tolerate coarse fodder, but the growth rate for the fattening pig is not satisfactory. We use yorkshire stud pig for matching and use their first generation hybrid to produce fattening hogs with a good result.

As people's living standard has been improving, more lean meat is required. Landrace hog for salted meat is being recommended for cross matching.

In the district where fodders are abundant, the first generation hybrid sow from the Yorkshire stud pig and Taihu sow is intercrossed with Landrace stud pig. The offsprings of this kind of triple hybrids grow very rapidly, and their hybrid vigor is more evident. In order to fully utilise the good stock, artificial insemination is often adopted for matching.

Fodders of the Pig and Rearing Management

On an integrated fish farm, there are various kinds of fodders which can be used as pig feeds, but we should have reasonable arrangements in order to increase their nutritional value, and promote the growth and fattening of the hogs.

In order to lower the cost of pig farming, local produce and by-products of the farm should be utilised to the maximum. The feeds should be complete in nutrition, including 55–60% of corn, barley, wheat bran and rice bran, 5% of soya bean cake and cotton-seed cake,

15–25% of chicken manure, 15–20% of wine lees and bean dregs, with an addition of 0.5% of table salt and 2–3% bone power and calcium carbonates. Green fodder is about 15–30% of the total fodder.

The following formula of the feed can be used in the area where is rich in fodder crops;

Weight of pig (kg)	***Piglet's fodder 5–20***	***Young pig's fodder 20–50***	***Fattening fodder 50–90***
Item			
corn (%)	53	50.5	45.66
Barley (%)	7.5	15	11
Sorghum (%)	12	5	11
Soya bean cake (%)	15	12	7
Wheat bran (%)	-	5	15
Fish meal (%)	10	5	3
Chinese scholartree leave powder (%)	-	5	5
Bone powder (%)	2	2	2
Table salts (%)	0.5	0.5	0.35
Total (%)	100	100	100
Additive to each ton of fodder			
Sodium selenite (g)	0.15	0.15	0.15
Zinc sulphate (g)	200	200	200
Potassium iodide (g)	1	1	1
Multivitamins (g)	40	40	40
Digestive energy (Cal/kg)	3,084	3,011	3,028
Crude protein (%)	19.3	16.41	14.32
Daily amount for each pig (kg)	0.3–1.2	2.1–2.2	2.2–3

The piglets may be the ones reproduced from the sow of the same farm, but usually the piglets are bought from the breeding farm.

Piglet fodder should be given to the newly bought piglings and young pig-fodder given after the body weight reaches over 20 kilograms. When the body weight reaches 30 kilograms or more, fattening fodders are maintained until the body weight is over 90 kilograms.

The piglets should be trained right upon arrival in the pigsty. They are trained to eat, sleep and discharge at a fixed place. The feeding trough is usually placed on the southern side of the hogpen, and a small amount of manure is put near the corner of the manure exit on the southern side in order that the newly-arrived piglets smell the excreta, and then have a habit of discharging at that place.

Clean grass bedding is stacked on higher place on the northern side for sleeping. Sometimes, piglets discharge everywhere in the hogpen, the stockman should sweep the excreta into a fixed corner to keep the sleeping place clean.

After several days, piglings could form a good habit and keep their sleeping place dry and clean. The piglets are fed three times a day with a fixed feeding time and amount. Hoghouse needs to be cleaned 2–3 times every day to make the piglings sleep and eat well.

Prevention and Treatment of Common Disease

The resistance of pig is stronger than that of chicken. If we can take good care of their health, including proper management, sanitary environment and preventive measure, pigs are not easy to get sick.

The common infectious diseases of pigs are: pig pest, pig erysipelas, pig pasteurellosis, piglet parathyphoid, and pig astheme. There are vaccines for the former four diseases, and "triple vaccine" for these three fulmingating infectious diseases, pig pest, pig erysipelas and pig pasteurellosis. It is convenient to use and very effective.

There has been no vaccine for pig asthema so far. The main measures are quarantine, eliminating diseased pigs and raising healthy pigs. Intramuscular injection of kanamycin, 20,000– 40,000 IU/kg of body weight once a day for 5 consecutive days or intramuscular injection of oxytetracycline, 20–40 mg/kg of body weight once a day for 5–7 consecutive days will have certain curative effects. The sick pig should be isolated, sterilised and dead pigs should be deeply buried or burned in order to stop the spreading of the disease.

Milk Cow Production

Cow is a herbivorous ruminant animal, and it can better utilise green forage, especially the chicken manure including grass bedding. Chicken manure contains a large amount of non-protein nitrogen compounds which exist in a state of uric acid or amide and can not be effectively absorbed by the pigs, but milk cow as a ruminant animal, can well utilise them. That's why milk cow is one of animals raised in "fish-livestock-and-poultry" integration. The milk output of the cow is high, economic benefit great, and excreta considerable. The main approach of raising milk cow is grazing.

So a pasture is needed and stall feeding is practised in winter. Because of lack of large pasture, intensive stall feeding is adopted for raising milk cow on fish farms. In front of double-rowed cowbarn, a play ground is enclosed for cows. They are fed and milked in the stall, and walk around in the play ground. All the forages needed by the milk cow are supplied during stall breeding.

Black and White Cows Suitable for Fish Farm to Raise

Black and white cow is the highest milk yielding species. Most of the countries in the world raise this kind of milk cow, which originated in the Netherland, but they all had bred their own species named after their own countries such as, American black and white cow, Japanese black and white cow etc.

Chinese black and white cow is one of the main species of our milk cows which can acclimate the natural conditions of the local district, and with high productivity of milk.

The primiparous cow could yield 4,000 kilograms of milk during 305 days of lactation, and over 5,000 kilograms of milk can be expected after third pregnancies. The milk fat index is 3.2– 3.5%. The Chinese black and white cow has high productivity, strong body and good adaptation, therefore, it is a good species for raising.

Establishing Sound Cow Groups

Tuberculosis and brucellosis are common infectious diseases for both man and cow. The characteristics of tuberculosis is to cause caseous degeneration of the tubercle nodes on the lung and lymph nodes while brucellosis is to cause abortion, infertility and decrease of lactation of cow.

These two diseases will influence both milk cow and human being, therefore, strict quaratine : measures should be taken in case a fish farm wants to develop dairy enterprise. Newly-bought milk cow should be isolated for 3 months and tuberculin tests should be performed thrice, by both intradermal inoculation and intraocular instillation. They can not mix with other cows until we are sure that they are disease free. Quarantine test should be performed twice yearly for those cows without tuberculosis. If there is any cow getting tuberculosis, it should be either isolated immediately or checked out. In that case, all the other cows should be examined again 30–45 days later until there is no positive case found for thrice.

All the workers in the dairy should be regularly examined by X-ray. If any tuberculosis case is found, the patient should be transferred out from the dairy. If calf is born from a diseased cow, it should be fed separately.

Enhance the breeding of healthy calf to replace diseased cow and thus, new healthy cow group can be established. Serum agglutination test can be used for the quarantine control of brucellosis of milk cow. It should be performed yearly.

Healthy cow uses sheep type No. 5 brucellus bacillus (or pig type No. 2) attenuated vaccine for prevention, but they have some bad effect on the quarantine test.

Forage and Feeding of Milk Cow

Stall feeding is normally carried out in the dairy on fish farms. The feeding of milk cow should be performed on the scientific basis; therefore, the formula of daily feeds and feeding plan must be worked out according to the standard which includes nutrient components of the forage and nutritional requirements of the animal.

These standards vary in different countries, each has its own general standard. For instance, U.S.A. has NRC system.

Feeding Standard for Milk Cow (Daily)

Table: Oat Unit Standard

Body weight (kg)	*Oat forage unit (kg)*	*Digestible crude protein (kg)*	*Ca (g)*	*P (g)*
350	3.7	210	18	9
400	4.0	230	20	10
450	4.2	240	23	12
500	4.6	260	25	13
550	4.9	280	28	14
600	5.1	300	30	15
650	5.4	310	33	17

Table: Milk Net Energy Unit Standard (NND)

Body weight (kg)	*Dry materials (kg)*	*NND*	*Digestible crude protein (g)*	*Ca (g)*	*P (g)*	*Carotene (g)*	*Vitamin A (I.U.)*
350	5.04	9.17	227	21	16	37	15
400	5.57	10.13	250	24	18	42	17
450	6.09	11.07	274	27	20	48	19
500	6.58	11.97	296	30	22	53	21
550	7.08	12.88	318	33	24	53	23
600	7.55	13.73	339	36	27	64	26
650	8.02	14.59	360	38	30	69	28

These are the minimum standards, and 10–15% can be added in practice. As the body weight of the primiparous lactating cow is still increasing, their standard should be 20% higher.

If the proposed milk production is over 6,000 kilograms, usually they are fed with the feed for the weaned multiparous cow with daily milk production of 15 kilograms.

Daily portion of forage for the milk cow is given according to the milk production (the amount of milk produced and its fat content index). The feeding standard can be calculated out according to the nutritional requirements of daily milk production plus the average body weight of the milk cow (counted as 600 kilograms). Then take the nutritional requirements of the feeding standard for milk cow for reference, calculate out the total daily nutritional requirements.

The physiological characteristics of different stages of the milk cow should also be considered in the calculation, e.g., pregnancy, etc. The price of the feeds and palatableness should also be considered. The forage of the milk cow is mainly green and coarse feeds. In winter or hay season, cows are fed on one kilogram of dry grass and stalk of the crops, 3–4 kilograms of silage or 5–7 kilograms of root tubers for every 100 kilograms of body weight. 8–10 kilograms of green grass are given to every 100 kilograms of body weight in grass growing season. One kilograms of mixed fine forage is given for production of every 2.5–3 kilograms of milk.

The ratio of fine forage mixture is 30% cake type feedstuffs (if chicken manure are used, only 15% can be added), grain feedstuffs (barley, corn) about 40%, bran 10–15%, by-products of processing 10%, minerals, salts and fish meal 10%. The above ratio is only for reference. The nutritional components of different forages can be calculated from the nutrients table, and also the amount should be adjusted according to the needs of the milk cow. The production of mixed forage developed very quickly in the recent two decades. Forage manufacturers use computors to calculate the needs of the milk cow, and mix the feeds automatically. They can produce and supply mixed whole forage.

Milk cow is fed three times daily with fixed amount at fixed time, adopting the method of small amount but frequent supply during feeding, usually fine forage given first, followed by fodders, and then water. The forage must be fresh. Iron nails and wires must be removed if there is any in it. Milk production can increase by 10–15% if sufficient amount of water is drunk, by milk cow. They should have sufficient exercises. The whole body of the cow is brushed before milking, dry brushing in winter, and washing and brushing in summer. Cow shed should be cleansed frequently in order to keep the shed and playground clean and sanitary. Cow manure and urine are washed directly through manure ditches into fishponds. 5–20% bleaching powder emulsion or 5% cresol is used for the sterilisation, which should be performed at least monthly.

Within 4–5 days after parturition, the milk in the breast of the high yielding milk cow should not all be squeezed out, only 2 kg of milk each time. One third of milk is milked out on the second day and then the amount could gradually be increased. All milk could be squeezed out on the 4th day. Dry fine grass, not forage with much juice and fine food, are given to the weak cow as its main food within 3 days after delivery. The milk production gradually increases 10–15 days after delivery, and reaches its peak of lactation.

Fine quality grass, sufficient forage and sufficient amount of water should be guaranteed at this stage to meet the nutritional demands of lactation. The milk production gradually decreases 3 months after delivery, and weaning starts 60 days before next delivery. 10 days before weaning, reduce the amount of fine feeds, green fodders and juicy forage and times of milkings as well. Do milking every other day or every 3–4 days. Stop milking when the production drops to 4–5 kilograms.

Milking Technique and Storage of Fresh Milk

Milking can be conducted manually or electrically. Integrated fish farms usually don't have large numbers of milk cow, therefore, manual milking is preferred. Wash the udder with 50°C warm water and then thoroughly massage the udder. Milking is performed by first massage of two nipples at a time. The first and second runs of milk are collected in a special container, and should not be mixed with normal milk in the milk pail. When a large part of the milk is milked out, the udder should be massaged again in all areas, and finally massage should be carried out the third time at the end of milking until all the milk is out.

The milk collected by manual milking should be filtered by gauze to remove hairs, dust, fecal materials and other impurities. It is to be cooled quickly and preserved in a cool place. The milk should be thoroughly sterilised and covered tightly and transported to milk-collecting station as quick as possible.

Artificial Breeding and Utilisation of Earthworm

Earthworm is good food for fowl and fish. Fresh earthworm contains 8–10% protein, while dry earthworm up to 56–66%. Its effective energy is 2920 Cal. per kg. Its nutritional value is equal to that of fish meal. The reproductive ability of earthworm is very strong, and it can multiply 200 times within one year under normal conditions. If good care is taken it can multiply 1000 times. No special equipment and fodders are required for the breeding. The fermented cow dung, pig manure, weeds and rank grass from the integrated fish farm mixed with proper amount of silt are good fodders for the earthworm.

Large-scale propagation of earthworm can not only improve the soil, but also increase the productivity of the crops and improve the environment. Artificial breeding of earthworm is now advocated and propagated in our country. Some of integrated fish farms take earthworm breeding as one of their animal raising. Earthworm can serve as protein feeds for both poultry and pigs or fish. The results are very promising in enhancing yields.

Plant Cultivation on an Integrated Fish Farm

Fertilizers and feeds are considered to be two of the basic conditions to strive for a high yield of fish. All fertilizers are used primarily to propagate natural organisms for fish to eat. Various kinds of forage grasses, beans, grains, melons and vegetables, aquatic plants, etc. are also good foods for fish. To establish an integrated fish farm needs certain amount of forage land and land for forestry, animal husbandry and side-line production as well, which are all directly beneficial to fish farming.

Crop production on an integrated fish farm means the utilisation of pond dykes, river banks, corner areas, etc. and the forage field. Cultivation of grains, beans, pasture grasses, melons and vegetables, fruit and mulberries should be rationally practised in accordance with the needs of fish and the growing seasons of different crops.

Furthermore, water surface at the edge of rivers & lakes in the vicinity of the fish farm can be utilised to cultivate aquatic plants so that high quality forage will be available at every season of the year. This kind of forage and silage can directly be used not only for breeding fish, but for rearing domestic fowls and animals or fertilizing water to grow plankton for fish.

Pasture Grasses

Perennial Ryograss (Lolium Pereme): It is a perennial plant belonging to the Gramineae family. It has such features as quick growth, high yield, rich in nutrition, easy to cultivate, low cost and strong adaptability. Its yield may reach 5–10 tons per mu. It is a good food for Grass carp, Chinese bream and Wuchang fish. Ryegrass cultivation can provide fresh food for fish in early spring, thereby, enabling them to break their fast ahead of season.

Seeding and Transplanting

Ryegrass is a tardy plant which grows even in the shade and likes moisture. Seeding is usually done at the end of September. Land should be prepared by applying 1,000–1,250 kg/mu of human feces as the base manure, or splash a layer of river silt over it, and upturn the soil by deep tilling, and then pulverise and smooth the soil. The amount of seed for broadcast sowing is 2–2.5 kg/mu. If the weather is dry after sowing, sprinkling is necessary and this should be continued till the complete emergence of seedlings. Then give another dressing of human feces of the same amount as the first application.

Transplantation is usually conducted at the end of October and early November. When seedlings grow to a height of 12–15 cm, they are pulled out to be transplanted in a field that has already been tilled and splashed with a layer of pond silt as base fertiliser. The leaves of seedlings should be cut half to ensure quick establishment. The spacing is 18 × 18 cm or 21 × 21 cm, and 6–7 seedlings are planted in one bunch. After transplanting, apply night soil in 1:3 dilution (night soil 1, water 3) to facilitate establishment and growth.

Field Management

Weeding: In order to enhance the growth of transplants, all weeds have to be eradicated. Weeding should be repeated now and then after transplanting.

Watering: Seedling plots and transplant fields should be kept moist and watering is necessary in prolonged fine weather.

Fertilizer Treatment: Usually one application of nitrogenous fertilizers is given before the initial cropping. Later on, organic manures should be dressed at a rate of 250–300 kg/mu after every harvest. The result is better if soil could be loosened and fertilizer applied the very day when it is mowed. Some inorganic fertilizers can also be used.

Harvesting

When ryegrass grows to a height of 30–60cm, cut them near the ground to serve as fish feeds. At this time, this grass is tender and so every bit of it is consumed. Thus, the utilisation rate of feed is high. Besides, ryegrass grows fast and has a high power of tillering after cutting. Generally speaking, when air temperature is low from October to next Feb., the growth of ryegrass is retarded, and only one or two outtings could be performed. From March to May, harvesting can be done once every 20 days or so. Be sure that the second and third cuttings are made closely to the ground, leaving stubbles only 2–3 cm high so as to increase tillering, harvesting and promote its quality.

Reservation of Seeds

English ryegrass begins to shoot, bear ears, and flower in April, and seeds can be collected in early June. Since seeds do not ripen at the same time and will fall to the ground easily, so when the colour of the ears turns yellow, collection should be done in time to prevent seed loss. The yield is about 50 kg/mu. Plots reserved for seed production should not be cropped as green fodder.

Sudan Grass

Sudan grass is an annual plant belonging to the genus Sorglum and Ropogon of the Graminese family. It can thrive in soils of different fertility; however, the fertile clayish soil is better. The yielding is about 10 tons per mu. Sudan grass has such features as the ability to withstand dryness and fertileness, rapid regeneration, high reproductivity, strong adaptability, high yield, superior quality and easy cultivation. It is a high-yielding crop in summer and autumn and so is suitable for subsequent cultivation of ryegrass to provide high quality fresh food for Grass carp, Bream and Wuchang fish from spring right through autumn.

Seeding and Transplanting

Sudan grass likes damp environment but dreads frost. Seeding is usually done when soil temperature above 10°C. The optimum temperature for germination is 20–30°C. Strip drilling is suitable for Sudan grass which is primarily out for green fodder. Row spacing is 20–30 cm and the depth of seeding 3–4 cm. Then, it should be mulched with a layer of plant ash 1 cm thick. The amount of seeds used for dense drilling is about 2 kg/mu. Besides this, dibbling or nursery seeding can also be practised. Plant spacing for dibbles is 18 cm. Each bunch takes about 10 seeds. When young seedlings are 12 cm high, they are ready for transplanting. The plants are spaced at about 18 × 18 cm with 4–5 seedings each bunch.

Prior to tilling and sowing, apply a layer of pond silt over the field as base fertilizer to be upturned into the soil, and then level and smooth the soil.

Field Management

Emergence of seedlings is completed 7–8 days after seeding. Budding takes place in about 70–80 days, and flowering in 80–90 days. The growth period is 100–120 days. Weeding should be done regularly. In prolonged fine weather, sprinkling should be given, especially during the seedling period. It is better to loosen the soil and apply fertilizer after every cutting. The method of dressing and the amount of fertilizer used are the same as those for ryegrass.

Harvesting

When Sudan grass grows to a height of 60 cm, harvesting is feasible. The cutting should leave a stubble of about 10 cm high, otherwise, tillering, regeneration and the yield will be affected. From mid May to Aug., harvesting can be done every 15 days or so. Growth slows down around September and death comes with the appearance of frost.

Bunch Grass Symphytum Peregroinum

This species is a perennial plant belonging to the Borraginaceae family. The region of origin is Iran-Transcaucasic. It is usually set in early spring yielding 7–9 tons per mu in the first year to be followed by 10–12 tons in the second year, and it may reach 15 tons. Fresh stems and leaves of this grass are tender and juicy, nutritious with high protein content. According to calculations from dry matter, it contains 22–25% crude protein, 4–6% crude fat, 7–13% crude fibre and 38–40% non-nitrogenous extracts. So it is an excellent green fodder rich in protein.

This grass has a well-developed root system with vigorous growth, and has strong adaptability to environmental conditions. It is susceptible to few pests and is easy to grow and propagate with a high survival rate. Its underground portion can withstand adverse temperature conditions ranging from -40°C to 37.8°C. Besides, it is resistant to draught and disease.

Seedling Cultivation and Setting

The organs of this plant above the ground like leaves, pedicles, rhizome buds, and those under the soil like rhizomes and roots can all be used as asexual materials for seedling culture and propagation. The methods of propagation in practice are division of suckers, root cutting, longitudinal cutting of root necks, setting buds divided from root and setting of stalk cuttings, etc. The period for nursery bed breeding is 30–40 days. When seedlings grow to a height of 10–15 cm, they can be transplanted and set at a spacing of 45 × 45 cm with about 2500–2600 plants per mu. In poor soils, increasing the plant population to about 3000 per mu is advisable. In dibbling, bury the plant up to the neck of the root which is the proper depth of setting. After that it should be sufficiently watered.

This grass is tolerant of a wide range of soil conditions, but a thick layer of sandy soil with good drainage is considered to be the most suitable. Plants can be cultivated on spare spaces around a pigsty and on slopes of pond dykes. This would facilitate to management. Land should be deeply tilled, harrowed, smoothed and levelled before setting the plants. It should also be well manured. It plays where is inconvenient broad ridgor with grcoves crossing each other will enable rain water to be drained off readily in rainy season.

Field Management

Dressing should be done once before the grass turns green in early spring. The amount of pig manure applied is roughly 2500 kg per mu. To the seedlings that are just set, measures should be taken to loosen soil and weed the field in time. After establishment, dress again to invigorate the settings and promote the development of roots. During the growing period, fertilise the soil each time after harvest, and also before overwintering. The last dressing will ensure stout plants to tide over the cold months. When the weather is dry and hot, sprinkling should be done timely while attention ought to be paid to draining in the rainy season. Water logging will cause the root to rot. This grass is quite resistant to pests and no serious case has been noted so far. However, in its seedling stage, cutworm attacks often occur and measures should be taken to control it.

Harvesting

When this plant begins to bud and flower in early and mid may, harvesting can be done. Cut the stems above the ground, leaving stubbles of about 6 cm to germinate anew. After that, harvesting is done at intervals of about 35 days. Thus, a year gives about 5 or 6 cuttings. In general, one planting would do for 8–10 years. Upon arrival of autumn, this grass will no longer bud and flower. The proper time for harvesting depends on the amount and colour of the foliage. Usually, no more cropping is done towards late October so as to retain the year's final coat of leaves for over-wintering. The crop is highly resistant to cold, and growth ceases only with the coming of frost which makes it gradually wither and die. However, the roots under the soil can even survive severe winter temperature as low as -40°C. In order to make a full use of land, inter-cropping can be done after auturn for other fodder crops.

Beans

Soybean (Blycine Max) (G. Soja) (Soja Max): The soybean is an annual plant belonging to the subfamily Papilinaceae of the Leguminousae family, the native plant of China.

The yield of soybean is generally 80–130 kg/mu. Soybean seeds contain 30–40% protein, 20–24% fat and about 30% carbohydrates. Besides these, there are several kinds of minerals and vitamins, and its nutritive value is very high. When ground and fed to fish fry, it forms an excellent food.

From seeding to maturity, the life of a soybean plant undergoes a few growth stages like the germination-seedling stage, the third trueleaf stage. The whole growing period is about 120–140 days. The length of time from germination to blossoming varies with variety.

Generally speaking, that for the early ripening varieties is 34–43 days, that for intermediate varieties 40–60 days, that for late varieties 50–60 days and that for very late varieties 63–84 days. They are tolerant to a wide range of soil conditions but soil layers that are thick and rich in calcium and humus with good drainage (pH value 5–8) are more suitable. However, they will not grow in soils where pH value exceeds 9.6 or is below 3.9.

Seeding

When the seed of soybean absorbs water of 100–150% of its own dry weight, it will commence to germinate if air temperature is 10–20°C, and at the same time, there is sufficient supply of oxygen. Under suitable soil moisture conditions, the seed will absorb enough water within one day. Seedlings will grow rapidly when air temperature averages 20–25°C in the day time and not below 15–17°C at night. Seeding time for spring and summer soybeans is late March to early April and June 1 to 20 respectively.

The amount of seeds required for seeding is 6–7.5 kg/mu. Seeds should be selected by winnowing, screening and choosing prior to seeding. The emergence of soybean cotyledons is relatively difficult, and so it is necessary to keep the soil loose. Besides, the germination of soybean requires ample moisture. Before seeding, the soil should be upturned and smoothed to preserve soil moisture and fertilizer. The method of seeding adopted is dibbling. Commonly, plant spacing used is 15 × 18 cm or 18 × 18 cm. One mu may have 15,000–20,000 dibbles, each of which takes 2–3 seeds. After seeding, they are mulched with a thin layer of plant ash or fine soil to facilitate emergence.

Field Management

Fertilizer Treatment: For every 50 kg of soybean produced, it is necessary to apply 2.8 kg of nitrogen, 5 kg of phosphoric acid, and 6.5 kg of potassium oxide. At the initial stage when nodules have not yet been formed at the root portion of the seedling or the action of the root nodules are rather weak, growth is slow. In order to raise soybean production, it is necessary to apply an appropriate amount of nitrogenous fertilizer and a large amount of phosphorus and potassium. In addition to applying sufficient base manure, seed-mulching manure is also feasible, i.e., to use a small amount of high quality fertilizer like decomposed barnyard manure, or high quality compost mixed with plant ash, or superphosphate to mulch the seeds seeded. Later on, by judging growth conditions, apply an appropriate amount of ammonium sulphate at a rate of 2.5–5 kg/mu during the seedling stage and 7.5–10 kg/mu during the flowering stage.

Checking, replanting and thinning for evenness: when soybean seedlings have uniformly emerged, a check should be made. Dibbles without seedling have to be replanted. In order to enable seedlings to grow stoutly and evenly, thinning should be done when a pair of true leaves have appeared.

Inter-row cultivation for weeding in combination with hilling: As is common, after seedlings are established, shallow hoeing for weeds in conjunction with thinning should be done thrice. From flowering to the swelling up of pods, soya bean plants are in utter need of water. If draught is to continue, serious reduction of yield will result. In places where conditions permit, rational irrigation will increase production to an appreciable extent. At the stage when pods are ripening, pulling all weeds out will better inter-plant ventilation and the penetration of sunlight into the foliage cover, thereby enhancing earlier ripening and ensure a bumper crop. At the same time, the work of prevention of such pests as the soybean aphids and pod borers, and bean hawkmoths, and the weed, dodder Cuscuta (love vine), should not be overlooked.

Harvesting

The mature stage for spring & summer-sown soybean crops fall on mid or late July, and early or Mid October respectively. When two-thirds of the foliage have turned yellow and fallen and the majority of the stems and pods appear dark brown with seeds separated from the inner walls of the pods, when the beans are already half-dried and quite hard and are of their original colour, and when you're shaking the plant some rattling sound comes from the pods, It can be ascertained that the time of harvesting has come.

All attempts should be made to gather in the crop within a week's time. Transport the harvest to the drying yard where it should be turned over and over to be sunned for a few days until it is thoroughly dry, and then separated from the pods on fine weather. After tossing the beans into the air to remove all impurities, and drying, it could be stored when their water content is below 13.5%, while those used as seeds the water content should be below 12% to ensure long storage.

Grains—Barley

Barley is an annual plant belonging to the genus Hordeum of the Gramineae family.

The yield of barley is about 200–300 kg/mu. Apart from being a food grain, barley is an important material used for brewing boer, producing alcohol and making maltose. Its straw is good for weaving The growing period for barley is about 190 days. Its lowest germination temperature is 1–2°C. The optimum temperature for growth is around 20°C. It bears ears in early and mid April with a milk stage of about 10 days to be normally followed by a wary ripe stage of 6–7 days.

Seeding

Seeding of barley is generally done in late October to mid November at a rate of 15–18 kg of seeds per mu. It can be broadcast or strip drilled. Its root system is mainly distributed within 15–18 cm of the cultivated surface layer. So secondary tillage should be done carefully with the "3-ditch" pattern of drainage. Seeding should be shallow and even. For producing every 50 kg of barley, it is necessary to fertilize the soil with 1.5 kg of nitrogen, 0.62 kg of phosphorus and 1.15 kg of potassium. Base manure to be applied is 6,000–7,500 kg/mu of grassdecomposed pond silt or pigsty manure 2,500–3,000 kg/mu. For the intarmediate soil layer, 15–20 kg of ammonium sulphate and 25 kg of calciua superphosphate per mu should be used with 3,000–4,000 kg of seed-Mulching manure per mu.

Field Management

Early stage field management: If seeding is followed by draught, measures should be taken timely to ensure total emergence of seeds evenly, and fertilizer be applied at the first-leaf stage, using 500 kg of night soil diluted with 3,000–3,500 kg of water per mu, or 10–12.5 kg of ammonium sulphate diluted with water for splashing over the surface. Another way is to use mainly organic manure made up of about 5,000 kg diluted silt mixed with 30–40 kg of dissolved ammonia water per mu, or applying 1,500 kg of pigsty waste and then splash a layer of thin river silt weighting 5,000 kg. Inter-row cultivation may be done 2 to 3 times. Cholraluron may be used for weeding on barley fields.

Intermediate Stage Field Management: Be prepared for application of fertilizer for reviving of seedling. In general, this is done by applying 7.5–10 kg of ammonium sulphate per mu in early February, or 20 kg of ammonia water dilution or 1,000 kg of night soil for splashing. Ditches have to be cleared to better drainage and prevent dampness. Towards the end of the month, be sure to make a dressing of about 12.5 kg of ammonium sulphate per mu to promote growth and the bearing of ears.

Late Stage Management: Be sure to do the work of disease prevention and pest control. Spray 90% trichlorfon in 1:1000 dilution to control army worms; 40% dimethoate emulsion 1:2000–3000 for aphids, and 0.2– -0.3 kg of carbendazol diluted with 75–100 kg of water for red mold and powdery mildew.

Harvesting

Barley is ripe by mid or late May. Harvesting should be done at the waxy ripe stage.

Corn (Zea Mays)

Corn is an annual plant belonging to the genus Zea of the Gramineae family. Its nutritive value is very high. The yield is generally 250–300 kg/mu and so it is one of high-yielding crops and a good fodder. Its grain contains 8.5% protein, 4.3% fat and 73% carbohydrates. So 50 kg of corn is equivalent to 67.5 kg of oats, or 60 kg of sorghum or 65 kg of barley in nutrition.

The nutritive value of corn stalk is usually more than double that of stalks or stems of other crops. Fresh stems and leaves harvested after the shooting of the phalanxes or before waxy ripe are green and juicy. Their nutritive value is very high. No matter whether they are used as green fodder or silage, they are good substitutes for fine feeds. Therefore, the successful cultivation of corn plays an important role in developing livestock farming.

Corn Thrives well in Warm Climates: The lowest temperature required for its germination is 10–12°C. It is a crop that needs short solar irradiance. In its young seedling stage, the amount of water it requires occupies 22.7% of the total needed for growth; the middle stage of growth calls for 44.5% while the late stage 32.8%. Soil suitable for its cultivation should be rich in organic material and well-drained, or of the sandy loam type. For producing every 50 kg of corn grain, the crop would take up 1.24—1.95 kg of nitrogen, 0.68–0.91 kg of phosphorus, and 1.96–2.07 kg of potassium from the soil. According to the length of its growing period, the varieties of corn to be planted can be classified into:early varieties (80–95 days), intermediate varieties (96–115 days) and late varieties (116–150 days). According to the sowing season, corn varieties are divided into spring corn, summer corn, autumn corn and winter corn.

Seeding

Before seeding, careful selection of seeds, examination of their germination rate and seed preparation through sunning, immersing and mixing with chemicals should be done in order to achieve the full germination, evenness of growth and get seedlings of vigorous growth. Spring corn can be seeded when soil temperature is above 10–12°C. To enable seedlings to have a longer period of growth, attempts should be made to sow the seeds as early as possible. Summer-and-autumnsown corn should be seeded as soon as the preceding crop is harvested. Seeding is usually done by drilling or dibbling. Seeding rate for drilling is 2.5–4 kg per mu; for dibbling 2–3.5 kg. Seeding depth is 3–5 cm.

Seeding density for late varieties is 2,000–3,000 plants per mu, for the intermediate 2,500–4,000 and for the early ones 3,000–5,000. When it's cultivated for fodder (silage or fresh), the density of planting may be heavier by about 20% compared with those just mentioned above. There are many ways of planting corn. Tilling depth can be 21–24 cm, and ample organic manures should be applied before land is tilled. In places where ridge cultivation is practised, special action should be taken to prevent draught and to preserve moisture for the top soil. In water logging or poorly drained areas, bedding cultivation should be adopted.

Application of base manure serves as principal source of nutrition for plants. Barnyard manure, livestock manure and compost prepared from the decomposition of stems and stalks which are all organic manures rich in nitrogen, phosphorus and potassium, are the best base fertilizer for corn.

Field Management

After the emergence of the seedlings from the soil, thinning should be done timely. Get rid of the weak seedlings and leave the strong ones so as to ensure their evenness and vitality. Quick-acting fertilizers such as night soil, ammonium, sulphate, urea, potassium sulphate and plant ash are applied as top dressing to promote growth during the three stages - the shooting up of stalks, the impregnation of ears and phalanxing.

Harvesting

After pollination, corn will be fully ripe in about 50–65 days. The higher the air temperature, the faster it will ripen. The leaves on the stem will turn yellow, the wrappers of the corn-cob will dry and wither, and the grains on the cob will look bright when it is fully ripe. It indicates the harvesting time comes.

Sweet Potatoes (Ipomoea Batatas)

Sweet potato is a trailing plant belonging to the Convovulaceae family. The region of origin is tropical America and Caribbean. Its yield is about 1,500– -2,000 kg/mu. The tuber contains 20–27% starch, 2.3% protein, 0.2% fat and also multi-vitamins. Besides its use as food, it is also a fine high-yielding forage crop.

Its tubers, vines and all residues after processing can be used as fodder. Its tender and juicy vines and leaves form a good fresh fodder. Its value as a feed is higher than ordinary forage grass. On one hand, sweet potato tubers can be used for feeding pigs and yet on the other hand, pig manure can be used to fertilize pond water. At the same time, its vines and leaves can be cut twice during July and August, yielding 1,000–1,500 kg of fresh vines which can also be served as fresh fish feeds. From the setting of cuttings to harvesting, the growth period for sweet potato plants is usually 110–160 days. Early tuber-bearing varieties can give a harvest in 80–90 days. The plant likes hot weather but cannot withstand cold.

It will stop growth at a temperature of 15°C and stored tubers will be "frostbitten" even at 9°C. It is a relatively draught resistant crop and will thrive in regions with an annual precipitation of 400 mm.

Seeding and Transplanting

In sweet potato production, most people adopt the tuber-vine-cutting method of propagation. Yield increase is directly proportional to the time the cuttings are set in advance. Spring cuttings should be set from late April to mid May and summer ones in June.

The density of settings is to be determined by such factors as climate, soil, the amount of fertilizers applied, the characteristics of the sweet potato variety, the length of the growing season and the method of cultivation.

As the sweet potato is a tuber crop, it needs a deep, loose soil layer for growth. Therefore, land used ought to be ploughed to a depth of 12–15 cm and then banked in ridges for setting the cuttings. To produce every 500 kg of tubers, the vines have to absorb 2 kg of nitrogen, 0.503 kg of phosphorus and 3.1 kg of potassium. The most suitable fertilizer for this crop is a compound fertilizer that contains N.P.K. such as barnyard manures and the compost. The manures that have more potassium like plant ash will give good, conspicuous results in increasing the yield.

Field Management

When roots of sweet potato cuttings are established, a check should be made to have the missing settings replanted. Inter-row cultivation and weeding should be done timely. After dressing and spraying, it is necessary to do tilling and weeding. In principle, inter-row cultivation ought not to damage the root system of the plant. Earth banking is not only of help in preserving soil moisture and preventing the tuber from being exposed, but also in draining water away as the grooves are deepened in the process of ridge banking which is usually done twice during the growth of the vines.

Harvesting

Harvesting comes when the weather gets cold and frost is approaching; when leaves near the stock have fallen and others turning yellow; when tubers have sufficiently swollen and their water content has reduced. Spring-set crops are harvested at the end of September to early October, while summer-set ones in mid or late October.

Melons and Chinese Cabbage

Chinese Cabbage

Chinese cabbage are biennial vegetables belonging to the genus Brassica of the Cruciferae family. It is the native plant of China.

The nutritive value of Chinese cabbage is very high. According to analysis, the deeper the leaf colour, the higher its nutritive value is. Each 0.5 kilogramme of Chinese cabbage contains 5.5 g of protein, 0.5 g of fat, 10 g of sugar, 2 g of crude fibre, and 4 g of inorganic salt. Besides, there are vitamins, carotenes, and mineral salts. In general, the yield is 1,500– -2,500 kg/mu. It serves as fresh food of high quality for herbivorous fish as well as domestic animals and fowls.

The optimal temperature for the growth of Chinese cabbage is 15–20°C. For seed germination, the optimum temperature is 20–25°C. A proper soil moisture should be maintained. When seedlings are young, nutrition requirements are small. In the middle stage when growth is vigorous, the demand for nutrients increases. Nitrogen is the major one it needs. The soil should be fertile with good moisture retaining properties.

Seeding and Transplanting

Spring Seeding: This refers to sowing and transplanting from the beginning of February to the end of May. Chinese cabbage that are directly seeded usually begin cultivation in early February.

Summer Seeding: This refers to sowing and transplanting from early June to the end of August.

Autumn Seeding: This refers to cultivating from early September to the end of October.

Winter Seeding: This refers to cultivating from early November to the end of next January.

Seed Rate: Varieties sown in early spring need 3.5–5 kg of seeds per mu; spring-sown ones usually 1.5–2.5 kg/mu; summer-sown Pakchoi (Brassica chinensis) 2.5–3 kg/mu; summer variety (direct seeding) 0.5–0.75 kg/mu; autumn-sown and transplanted 0.75–1.0 kg/mu; and winter nursery bed-cultivated 2.5–3 kg/mu.

Method of Seeding: Chinese cabbage seeds are sown by broadcast. Thinning, weeding, manuring, spraying and pests control should be done after the emergence of seedlings.

Field Setting: Setting is also known as transplanting. Different cultivation seasons have different requirements on setting. Seedlings for transplantation before August are usually about 25 days old; those transplanted in September, 28–30 days old; those transplanted in October, 32–35 days old; and those for transplantation in November, 35–40 days old.

Field Management

The field management work includes top dressing, sprinkling/watering and disease/insect control.

Harvesting

The harvesting period of the Chinese cabbage differs a great deal as cultivation methods vary. Young Pakchoi seedlings directly sown in summer can be harvested in 20 days; those cultivated in winter, in 80– –90 days; those transplanted from mid September to the end of the month, in 25–30 days; those transplanted in October, in 40 days; and those transplanted from mid October to early November, in 45 days.

Cabbage (Brassica Aleracea cv. Capitata)

Wild cabbage is a biennial plant belonging to the Cruciferae family. It is the native plant of southern Europe. The variety commonly cultivated in China is the ordinary cabbage. The yield per mu can reach 2,000–4,000 kg. It is rich in nutrition, having proteins, carbohydrates, mineral salts and vitamin C. Besides being edible, it is a very fine fodder for animals.

Wild cabbage, being a biennial just forms the edible globular top in the first year, and it is until the following spring when air temperature is still as low as 2–6°C that it starts to initiate flowering buds and then out to blossom and bear seeds. It adapts itself nicely with good resistance, through the optimum temperature for its growth is 14–20°C. It grows well in fertile soil that can retain moisture effectively with good drainage and easy irrigation. During the stage of formation of the globular top, it needs ample fertilizers and water. Nitrogen claims the largest share of its need though a considerable amount of phosphorus and potassium are also required. Its resistance to disease is good.

Seeding and Transplanting

Cabbage can be planted all the year round. According to harvesting season and cultivation method, it can be divided into three kinds: spring cabbage, summer cabbage and autumn-winter cabbage.

Field Management

Spring grown cabbage usually needs several applications of top dressing before the year ends. Night soil is dressed in 1:2 dilution and the amount to apply is about 1,500 kg/mu, one week after the setting of the seedlings. The second dressing should be given before Jan. 19 to enhance sprouting early in spring. The concentration and amount of manure are the same as the first dressing. The third one is to be applied one month later, and the concentration and amount to dress being 1:1 and 2,000 kg/mu respectively. The fourth dressing depends on the stage of growth, that is, when spring cabbage is forming its globular top in March to early April. The rate of diluted night soil applied is 1,500–2,000 kg/mu to accelerate the growth of the globular top and make it solid. It is necessary to dig drainage grooves in early spring. From the setting of the plants to the closing-up of ridges by their foliage, tilling is conducted in combination with weeding 3–4 times.

Summer grown cabbage (Brassica broccoli) usually need three dressings within a month. Each dressing calls for 1,500 kg/mu of human feces. The first dressing is given in 1:3 dilution after the set transplants have been established.

The second one is dressed 10 days later in 1:1 dilution. Before the formation of the globular top, the third is applied in 2:1 dilution. When the globular top begins to form, no more dressing should be done as it is liable to cause rotting, during the whole process of field growth, chemical fertilizers can also be applied singly like aqueous ammonia in 1:100 dilution 4–5 times at a rate of 25 kg/mu. The soil is generally loosened 2–3 times.

Autumn grown cabbage generally gets 3–4 dressings, each applying 1,500 kg/mu of night soil but in different concentrations. The first dressing is given 3–4 days after the transplants have been established in 1:3 dilution. Half a month later comes the second in 1:1 dilution. The third is applied before the formation of the globular top in 2:1 dilution.

The cabbage for over-wintering should not be dressed too heavily in the second and third dressings. In the middle of November, dressing is given to back up globular formation of the top using 2,000 kg/mu in 2:1 dilution.

Harvesting

Spring grown cabbage are harvested in batches sooner or later from mid April to early June depending on the variety. The yield is 2,500–5,000 kg/mu.

Summer varieties are seeded in early March and harvested in July-August with yields of 1,500–3,000 kg/mu. Those sown in early

May are gathered in August-September with yield of 1,500–2,000 kg/mu. Those seeded in early June are collected in September-October with yield of about 2,300 kg/mu.

Squash (C.maschata, Cucurbita Maxima, C. Melopepo)

The squash is an annual plant belonging to the Cucurbitaceae family. It is the native plant of the tropics.

The squash is a juicy fodder of high yield. It is rich in nutrition containing a lot of carotene as well as vitamin C and glucose. Its stems and leaves can be processed into stalk sugar fodder after drying. Its yield is generally 1,000–2,000 kg/mu.

The optimum temperature for the germination of squash seeds is 25–30°C while the most suitable one for the development of the fruit is 25–27°C. It belongs to the type of crops that need short solar irradiance. It thrives well in a dry, hot environment with a soil moisture of about 50%. A fertile, neutral or slightly acid sandy loam soil is best suited for its growth.

Seeding and Transplanting

Nursery bed seeding period usually falls between mid March and early April. Two kinds of seeding are in general practice: one is bunch planting, the size of planting holes is usually 7.5–9 cm square and 9–12 cm deep. Each hole takes an average of 2–3 seeds after which it is mulched with a layer of nutritive fine soil which ensures them not to be exposed to the air.

Spread a thin covering of straw over the bed to serve as shading and to preserve soil moisture, or place a few short bamboo sticks over it and then spread a plastic sheet on them to act as cover for the same effects. Finally, fit up a glass window frame and use mud to seal up the edges of the covering. At night use some straw mats to blanket the whole thing to prevent the loss of heat and accelerate germination.

The other method is "broadcast" sowing (also known as 2-stage breeding). Over the surface of a dressed bed is spread a layer of plant ash and then cast the seeds evenly on it. After that, use a tiny-hole nozzle sprayer to sprinkle fine droplets over it and then mulch the seeds with a layer of nutritive soil of 2 cm thick.

Finally spread a layer of straw over the bed or cover it with a plastic sheet by setting up a glass window for the sheet covering and then seal the nursery bed tight. At night cover it with a straw mat to preserve the heat in it to enhance germination. It's necessary to maintain proper temperature, better ventilation, sufficient sunlight and to control pests in the nursery.

Setting the transplants: When squash seedlings have grown to a height of 9–12 cm with 2–3 true leaves, they can be transplanted. They are usually set in late April, and intercropping with other vegetable crops.

Harvesting

Early set squash can be collected 10–15 days after the pistillate flowers have withered. Harvesting in batches at regular intervals commences from the end of July, and large area collection takes place in early and mid August.

Aquatic Plants

Water Peanut or Alligator Weed (Alternanthera Philoxeroides): The water peanut plant is a perenial plant which is a native plant of Latin America and was introduced into China more than 30 years ago. This plant is a high-yield, easily cultivated aquatic crop and good for both fodder and manure. Surface of lakes, river bends, ditches, and ponds can all be used for float cultivation.

It is not difficult to propagate, and relatively easy to over-winter with strong adaptability. The yield of fresh grass can reach 15–25 tons per mu annually - an amount enough to provide 10–15 pigs with silage all the year round.

Its stalks and leaves contain 2.49% non-nitrogenous extracts, 2.18% crude protein, 0.18 crude fat, 1.19% of crude fibre, 1.25% ash content, 0.232% calcium and 0.028% phosphorus. It can not only be served either fresh or cooked, but can also be prepared as fermented fodder or dried and ground as fodder for use throughout the year.

Water peanut plants love warm climate, humidity, and sunshine. They also need fertilizer. Although they are not strict about the depth of water on which they thrive, they can grow well in rivers, ditches and ponds which have slow running or stagnant water that is fertile with a depth of 1–1.5 meters. They undergo asexual reproduction through their stalks and vines.

As seedlings can germinate and grow on each and every node, they can propagate very fast.

Time for Breeding and Propagation

The growth and reproduction of water peanut plants are very fast. With the exception of the utter heat in summer and the bitter cold in winter, the rest time of the year are fit to make cuttings for transfer and propagation. However, if it is to be introduced into a locality for cultivation from another region, it is best to do so at the beginning of April when the old stalks are starting to germinate or just before they germinate. On the arrival of these introduced cuttings, they should be promptly cast onto the surfaces alloted to the within one or two days with the aim of promoting their early germination for earlier cultivation. Shelter them from wind and sunlight to avoid the risk of being dried. Self-reserved seedlings are best cut for cultivation after new stalks grow to a height of 15–30 cm. Cuttings can float better in this way, thereby, warding off the onslaught of waves.

Method of Planting

Generally, two methods are in practice: Laying ropes and using lattice frames. In places where the flow is little swifter and the water somewhat deeper, the method of laying ropes is adopted to prevent the seedlings from being scattered by the wind. While in stagnant ponds latticed frames are used.

Management

Fertilizer treatment: In general, water peanut needs no fertilizer. If water is sheer and leaves of seedlings turn yellow at the start or in the process of growing, then it is necessary to give top dressing such as fertile silt mixed with barnyard manure, or to spray diluted nitrogenous fertiliser on time in order to promote growth.

Weeding: This should be done before planting so as to wipe out all obstacles in growing water peanut. If weeds and moss reappear after cultivation, they ought to be timely wiped out. Moss is generally eradicated manually. If the results are not so good, it is better to spread plant ash or copper sulphate to kill it. This is done in two ways: One is to use 0.2–0.5% copper sulphate solution to kill the moss on them; the other is to crush copper sulphate into small pieces and put them in a small bag hanging on water surfaces densely aggregated by moss. With the gradual dissolution of copper sulphate in surrounding water, moss is killed.

Pest control: Three-spotted plusia occurs in July to September crawling over and devouring water peanut plants. To control them, use 90% crystalline trichlorfon at a concentration of 0.083 ppm for spraying, after which, 3–5 days have to elapse to allow the chemical to lose its toxicity enough before they can be harvested and fed to livestock. Steps can also be taken to remove a part of them away so that affected erect stalks will bend over and lie on the surface of the water, thereby drowning the pests.

Harvesting

After planting for 30–40 days with stalks and leaves rearing 20–30 cm above the surface, harvesting can be done, leaving stalks of about 6 cm above water and retaining about 3–4 leaves on them. After that, cuttings can be made every 10 days or so. In July and August, the weather gradually becomes warmer with abundant rainfall. Thus, water is enriched, the growth is fast and so harvesting at shorter intervals is advisable. After every cropping, use a hoe to loosen and separate the seedlings. This will help restore growth through better spacing. Towards the end of October comes the last harvest of the year.

Seedling Reservation and Over-wintering

Water peanut plants to be reserved for seeding should be densely cultivated on sunny leeward ponds. Choice should be made of plants whose stalks and leaves look luxuriant, relatively well-seasoned and healthy. Reserves should not be cut for harvest after Frost's Descent. Stalks and leaves of such seedlings should remain 30 cm above the water surface. Thus, when the leaves on the top are frost-hit and wither, they will note: Frost's Descent, 18th solar term, around Oct. 23. Naturally fall over and cover up the plants beneath them, thereby, serving as a protective layer which is of help in over-wintering the crop.

Water Lettuce (Pistia Stratiotes)

It is a wild aquatic plant belonging to Araceae family.

The roots of this plant look like a bundle of cotton threads suspended in the water: The stalk is very short and leaves grow in clusters. Each plant has 6–10 leaves arranged in a ring-like fashion. Leaf blades look oval in shape, on both surfaces of which grow velvety hairs. Its flowers are yellow, and the stamens without pedicles grow in symphysis but look protruded, while the pistils are like solitary bulbs of oval with single ovary in which are found several ovules. The fruit is a kind of berry.

Water lettuce plant contains 1.07% crude protein, 0.26% crude fat, and 1.63% carbohydrates. It also contains considerable amount of crude fibre and It can be used as fresh or cooked fodder for pigs. It grows and reproduces very rapidly and the yield is very high, 10–20 tons per mu. At the same time, it has such advantages as less labour, lower cost, no land-use and easy management.

Water lettuce floats on the water and so surfaces of river bends, lakes and ponds, etc. can all be used for propagation. It grows well in places where the water is fertile and stagnant. On running water, its growth is relatively poor. The suitable depth of water for cultivation is 0.7–1.5 m, and the optimum pH value is 6.5–7.5. This crop likes warm weather, and its resistance to cold is poorer than water peanut and water hyacinth, but its endurance to heat is better. Although it can grow in any temperature between 15–40°C, it grows faster when it is 22–35°C. It divides and reproduces the fastest, especially between 30–35°C. If water, fertilizer and sunlight conditions match nicely, each plant can reproduce 50–60 plants within a month. When temperature is above 35°C or below 18°C, it is unable to divide, or just divide a few times. When it is below 10–15°C, it will just maintain its life activities, and when it falls below 5°C, it will perish at once. Its endurance to fertility is very high, and so it demands an ample supply of nitrogenous fertilizer if it is to grow well. If water quality is poor and no fertilizer is applied, it will not thrive well.

Small Pond Planting in Spring

When air temperature rises above 15°C around mid April, small ponds can be used to plant seed seedlings removed from the nursery beds for preliminary propagation in order to supply sufficient seedlings for extensive planting in summer. When seed seedlings are transferred to ponds be sure that they are getting crowdedly together at one corner of the pond so that they will mutually give support to one another and not move about freely. Every measure has to be taken to prevent filamentous green algae from entangling the seedlings, and to provide such good conditions as fertilizer, air temperature and humidity, and places with still water for their propagation so that general cultivation may commence ahead of season.

Large-scale Planting in Summer

When the temperature of the water surface has risen to about 23°C around mid May, the growth and reproduction of water lettuce increases in speed conspicuously.

It is necessary to catch hold of this opportunity to expand the cultivation of this crop. The amount of seed seedlings to use on water surface of one mu is 10–20 kg. When planting, it is necessary to use reed stalks, bamboo poles, or tying straw ropes together to form an enclosure to cultivate them inside it so that they may thrive and grow in a colony to speed up reproduction.

If the population increase meets with a check, the enclosure should be enlarged gradually to provide more space for growth, and when the seedlings have occupied more than half of the pond surface, the ropes confining them can be removed to let the crop propagate at will. Later on, they can be transferred to other places for cultivation and expansion.

Water Surface Management

Fertilizer treatment: Fertilizer is applied 5–7 days after planting. After that, the dressing is done weekly, each time at a rate of 250–300 kg of night soil per mu or pigsty manure diluted 4- to 5-fold to be splashed onto the leaf surface in the evening.

Prevention and control of weeds: On the initial stage of planting, it is compulsory to eradicate all weeds on the water surface. If filamentous green algae is present, kill it by casting plant ash or spraying 1% lime water or 0.5% copper sulphate.

Pest control: Use 0.5% 666 powder to spray on the leaves of water lettuce early in the morning when their surface are still wet with dew to control aphids, or use 40% dimethoate emulsion in about 1:2000 dilution for spraying. At the initial stage of the occurrence of yellow wilt, use the Bordeauz mixture of 160- to 200-fold water suspension for spraying to control it. Crop can be harvested for fodder one week after applying chemicals to avoid poisoning of livestock.

Sprinkling water on leaf surface: In bright summer weather when the day is hot and dry, it is necessary to sprinkle clear water on leaf surfaces 2–3 times at noon to increase humidity and lower air temperature locally, thereby, promoting growth.

Harvesting

Cropping usually commences from mid or late June. First, divide the crop into several square blocks and then scoop up the plants block by block. The amount to gather depends on growth conditions. Be gentle in making the harvest in order to avoid too much shake-up of the crop. After harvesting, the seedlings should be brush apart so that they are distributed evenly on the water surface to enhance their propogation.

Seedling Protection and Over-wintering

From late Oct. to early or mid April, the work of protecting the seedling for over-wintering should not stop until it is entirely frost free. The work is done on the nursery beds. To maintain temperature properly is the prerequisite to the protection of seedlings. Therefore, it is essential to keep the temperature of the wintering nursery above 15°C. In addition, more sunlight and better ventilation will help the crop to overwinter safely.

Water Hyacinth (Eichhornia Crassipes)

It is a perenrial aquatic plant belonging to the Pontederiaceae family. The plant floats on the water, growing out creeping branches from its roots to form new branches. The leaves grow out straight and are either oval or round in shape varying from 2.5–5.0 cm in width. They are smooth and shiny. The petiole bulges out on the lower half like a gall bladder, a spongy interior filled with air. The flower is monopedicle with a sheath in the middle portion and flowers of six petals blossoming on the top in violet and blue. There is a bright yellow spot at the centre of the top petal.

The yield of water hyacinth is very high, reaching 10–16 tons per mu. It is also rich in nutrition. Analysis shows that fresh crops contain 1.9% crude protein, 0.25% crude fat, 1.11% fibre, 2.21% non-nitrogenous extracts. 1.33% ash content and inorganic salts like calcium and phosphorus. Therefore, the planting of water hyacinth is a good way of solving green forage for pigs. This crop likes warmth and humidity and still water or slowly running water. The optimum water depth is 0.3–1 m but the water has to be fertile. Its adaptability is better than water lettuce; especially its endurance to poor fertility is higher. The river bends and ponds that are not fit for planting water lettuce can be used for cultivating spring water hyacinth, and yet produce a considerable amount of forage. Besides this, it is more cold-resistant than water lettuce. As long as temperature remains at 7–10°C. it can safely overwinter.

Planting Time

Water hyacinth is cultivated along the Changjiang River Drainage in early or mid April. When air temperature rises above 13°C and frost is over, the old seedlings that have overwintered start to germinate and sprout. New leaves burst forth with all vigour. This indicates that it is the right time to plant hyacinth.

Method of Planting

Small stagnant ponds of less than 1 mu can be directly seeded with water hyacinth seedlings which may float and grow in it at will, while in large ones or in slightly running water, planting should be done with the help of bamboo frames or by setting up supports and then tying straw ropes to them to enclose an area of the water surface for planting the seedlings within it. When the enclosure is densely populated by them, enlarge it gradually until eventually the supports and ropes are removed. In this way, the crop will not be dispersed by winds and waves, and so it is advantageous to its growth and reproduction. One mu of water surface needs about 4–6 kg of seed seedlings.

Management

Fertilizer Treatment: In fertile ponds, no application of fertilizer is required, but it is necessary to stir the fertile silt at the bottom to enable nutrients to dissolve in the water. If water quality is poor, the plants will be weak and slender, and leaf blades will look yellow. It shows that night soil or animal manure need to be applied to the crop. The dressing is usually done in the vigorous growing season weekly or biweekly.

Weeding: During the initial stage, the growth of water hyacinth is easily interfered by weeds. So it is necessary to eradicate the aquatic weeds and filamentous green algae. Weeding will cease only when the growth of water hyacinth is vigorous and when seedlings cover up the whole water surface.

Cropping: When water hyacinth plants are growing luxuriantly and reproducing rapidly after planting for a month or two, harvesting can commence. The amount to gather is about one-fourth of the whole crop or one third at the most. After cropping, it is necessary to brush the seedlings apart for speedy reproduction. In summer, it is possible to gather the harvest once every week, but this interval of cropping depends on the fertility of the water, the growth condition of the crop and air temperature.

In Changjing River Drainage, the growth of this plant ceases by the end of October when temperature drops below 10°C and leaves begin to turn yellow. At this moment, seedlings that are of medium size, that look more seasoned and healthy and that are free from diseases and pests are to be chosen for overwintering.

The method involves placing a layer of fertile pond silt at the bottom of large vessels or wooden tubs and fill it with water to the 9 cm level. Then put the selected seedlings into it and place the containers in a nursery, the temperature of which is maintained above 7–10°C. The seedlings should be frequently exposed to sunlight and can be placed together with water lettuce seedlings in the same nursery for overwintering.

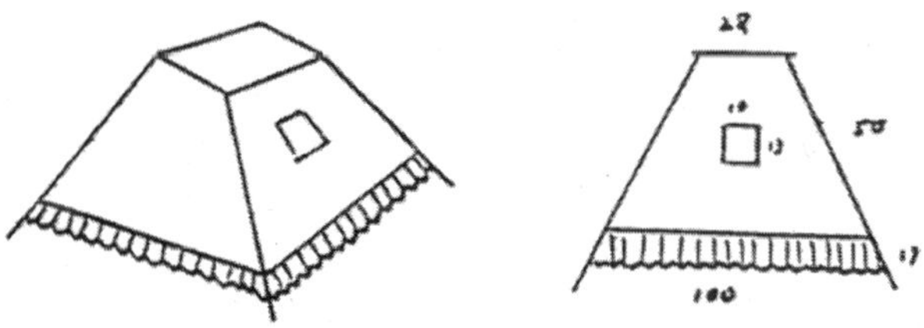

Figure : *Thermo-electric umbrella for chick nurture*

Ecological Efficiency of Pig-grass-fish Integration

To Raise Utilisation Efficiency of Light Energy

The products of photosynthesis depends on the amount of light that is the product of light intensity by sunshine time. Terrestrial pasture grasses can absorb light energy from the sun directly and can avoid shading each other under artificial control, e.g. mowing in time, so as to reduce the reduction rate of radiant energy. The rate of reduction of radiant energy in water is higher than that on land because the water surface reflects and scatters light.

When the sun rises high, the light intensity will be reduced 6% because of reflection of calm surface or 10% because of reflection of waving surface. When the sun rises or sets, most of light can not enter into water due to the shading of pond sykes, therefore, so far as the sun light is concerned, daylight in water is shorter than daylight on the land.

If the gradient of a pond dyke is 30 degree the sunshine time of the pond water along the pond dyke is reduced by one third in comparison with that on top of pond dykes. The lower the water level and the smaller the pond area, the shorter the sunshine time. Since there is a large amount of suspended and dissolved substances in fish ponds and especially plankton propagate greatly, auto-shading occurs and it's hard to control. According to the experiments of Freshwater Fisheries Research Centre, the transparency and the amount of suspended substances are in inverse proportion. In a natural pond, the coefficient of correlativity r=-0.9083. The regression equation is as follows:

$$Y = 247.692 - 2.840\ x$$

Y = transparency (cm)

X = the amount of suspended substances (mg/L)

Larcher (1980) realised that radiation intensity is in inverse proportion to the water depth, decreasing in the form of exponent (or index). When the sky is evenly clouded, the light intensity, according to synchronous measurements, on the soil surface with the depth of ryegrass of 40 cm, is 900 lx, which is 13.43% of that on the surface of grasses and that means its relative light intensity reaches 1.8% of 50 klx, the saturated point of net photosynthesis of land grasses; the light intensity at 40 cm below the water surface of a pond is 28 lx, the relative light intensity is only 0.42% of that on water surface. It only reaches 0.14% of 20 klx, the saturated point of net photosynthesis of phytoplankton.

Moreover, 40 cm is the maximum height of ryegrasses to be mowed. In the growing or ratooning phase before mowing, the relative light intensity is much higher than the measured one while the relative light intensity in the pond cannot be increased artificially. Therefore, from the view point of light intensity and sunshine time net photosynthesis of phytoplankton cannot compare with that of terrestrial plants either in a specific time or in a period of time.

Terrestrial grasses have greater ability to photosynthesize than phytoplankton. Under the normal supply of carbon dioxide and with an optimal light, temperature and water conditions, the uptake of carbon dioxide of sunplant herb is 20–50 mg/dm^2/hr. The intake of energy and the primary productivity of terrestrial grasses are higher than those of phytoplankton. Manured with pond silt alone, the yield of ryegrass can reach 4500 kg/mu in Apr. and May. The average daily output is 82 kg in May. The average daily yield of Sudan grass can reach more than 100 kg/mu.

The net output of oxygen produced by phytoplankton in Wuxi or Xiamen is 8–10g o_2/m^2/day. So the net yield of phytoplankton is only 42.3–52.9/ma/day. (Le xing-zhi et al). From the measurement in Wuxi high yielding fish pond, the net yield of phytoplankton in peak period was 41 kg in wet weight per mu per day, only 50% of the net yield of rye grass, 41% of the net yield of Sudan grass per mu per day.

After calculation, in accordance with the balanced equation of photosynthesis, 1 kg of O_2 is equal to 6.1 kg of phytoplankton in wet weight while 1 kg of O_2 means producing 3.51 Mcal of energy. Therefore, 1 kg of phytoplankton averagely holds 0.575 Mcal of energy. Daily output of plankton from May to Aug. is 41 kg/mu, that is, 23.575 Mcal. 1 kg of ryegrass holds 1.035 Mcal.

Daily output of ryegrass in May holds 84.87 Mcal, that is 3.6 times that of phytoplankton. 1 kg of Sudan grass holds 0.75 Mcal. Daily output from June to Sept. is 77.25 Mcal, which is 3.3 times that of phytoplankton. In the case of rye grass and Sudan grass. We can see that the primary productivity of terrectrial grasses is 2 to 4 times that of phytoplankton in high yielding fish ponds and the output of energy 3.3–3.6 times. The yield of terrestrial grasses can be further increased through cultivation and popularisation of good varieties of seeds, rotation systems and intercropping with leguminous grasses and scientific management. On the contrary, how to increase the primary productivity of phytoplankton is a hard nut to crack so far.

The utilisation rates of primary productivity of phytoplankton and terrestrial grasses by fish are different. Furthermore, phytoplankton can continue their propagation only after part of seeds remain in a pond. He Zhi-hui (1983) calculated the consumption of phytoplankton in high yielding fish ponds in Wuxi. The direct phytoplankton consumption of Silver carp, Bighead, Tilapia and Crucian carp or through zoopalnkton accounted for 82% of the biomass of phytoplankton. (70% of the zooplankton are consumed by plankton-eater.); whereas the utilisation rate of ryegrass with the height of 50 cm or Sudan grass below 1.2 meter by fish is 100%.

To Utilise Two Primary Productivity

Compared with manuring the pond water, pig-grass- fish integration utilises high effective primary productivity of terrestrial grasses and at the same time, it also uses the primary productivity in fish ponds. In this system, herbivorous fish are the major species, which are stocked according to the stocking pattern with net yield of 400 kg/mu. The mean capacity load of herbivorous spp. is 190 kg. The daily Sudan grass ingestion of herbivorous fish occupies 22.5–30% of the body weight of fish, average 25%. Thus, the ingestion of herbivorous fish is 48 kg/mu/day. The excreta of herbivorous fish are about half of its ingestion amount, so the daily amount of excreta are 24 kg. On the basis of analysis of South-Western Normal College in China (1977), the N and P contents of Grass carp excreta are 1.102% and 0.426% which are three times and two times those of pig excreta respectively.

At least, the daily excreta of herbivorous fish are equal to 24 kg of net pig manure. During the 250 days of ingestion period, the amount of excreta of herbivorous fish are equal to 6000 kg of net pig manure or the amount of net excreta of 6 pigs. It's near to the total amount of manure application in the fish pond with Silver carp and Bighead as major spp. Thus, growing grass by using pig manure can provide sufficient fodder grasses to the herbivorous fish, the excreta of which, in turn, can supply nutritive salts enough for the fish pond so that the primary productivity of the fish pond can reach the mean level of the pig-manured pond.

In Wuxian, Zhangzhuan Aquaculture Brigade which has adopted pig-grass-fish integration, all the pig manure are used to grow grasses. No additional manure is applied to the fish pond with herbivorous fish as major spp., which occupy 55–72% of the total stocking amount. The water colour looks green and brownish green with transparency of 10–35 cm. For example, in pond No. 33 with herbivorous fish occupying 72%, the average transparency is 17.5 cm. According to the regression equation of phytoplankton and transparency after He zhi-hui.

$$Y = 153.69 - 3.06\,x$$

Y = the biomass of phytoplankton (mg/l)

X = transparency (cm)

The biomass, of phytoplankton of that pond is about 100 mg/l. The biomass of phytoplankton of high yielding fish ponds in Wuxi is between 20 and 100 mg/l. It indicates that the excreta of herbivorous fish can fertilize the pond water and provide sufficient biomass of plankton demanded by plankton-eater.

Pig-grass-fish integration can avoid certain ecological defects caused by direct application of pig manure in ponds. The direct application increases the BOD and lowers the dissolved oxygen levels esp. if stacking application of manure is adopted at ten-day intervals. The BOD_5 = 29.09 – 30.75 g/kg.

According to the experiments of Fresh water Fisheries Research Centre, the average BOD_1 of fish ponds are 4.10 mg/l and 5.48/1 respectively when pig manure is spread evenly into the pond at the rate of 0.12 kg and 0.29 kg/m^3/week. At 5:30 in the morning, the average D.O. is only 1.0 mg/l. Zhangzhuan Aquaculture Brigade adopted pig-grass-fish integration in 10 ponds, and supplement molluscs and fine feeds without any other manures in fish ponds. The carrying capacity of 9 ponds were from 779.3 to 1350.5 kg/mu except one pond with 475.3 kg/mu. All Dissolved oxygen levels were high. The minimum DOL was 2 mg/l and average 4.6 mg/l in Pond No. 10 with the highest carrying capacity.

Because this system utilises two primary productivity, avoiding some ecological defects, there is a greater potentiality of increasing fish yields. In terms of many reports from abroad, the fish yields of filterfeeding species and omnivorous species in a manured pond only reach 15–31 kg/ha/day when they are monocultured or polycultured as major species. Generally speaking, the yield of plankton-eater is 300–400 kg/mu. It's difficult to surpass this figures and moreover, the economic efficiency will be reduced when the yield increases. Nowadays, in most of fish ponds with the yield of 500 kg, the herbivorous fish are cultured as major species with grass as main food and without application of manure or with just a little manure.

The excreta of 1 kg of herbivorous fish can fertilize the pond water and produce 0.2–0.5 kg of Silver carp and Bighead. The larger the size of fingerlings the more their ingestion, and the more the number of planktoneater that they bring along. In Grass carp rearing pond with gross yield of 110–150 kg compared with Grass carp rearing pond with gross yield of 50–100 kg, the net yield of plankton eater is 37.5 kg higher.

In fish-pig link, filter-feeding fish and omnivorous fish feed on detritus and the bacteria on detritus. The relationship between the fish yield and both detritus and bacteria depend on their quality rather than on their quantity, therefore, it can not be overestimated just according to the amount of detritus in a water body. The experiment of Lin Wan-lian also proves on one aspect that it is a loss to use pig manure to grow grasses for fish farming so far as part of detritus, which can be utilised by fish, is concerned. Nevertheless, the good effects of pig-grass-fish integration can make up for the loss. What's more, the effects of detritus is not so good.

The Economic Efficiency of Pig-grass-fish Integration

On the basis of ecological principles and its good economic efficiency, the pig-grass-fish integration can get better economic efficiency than fish-cum-pig integration. In 1979, Zhangzhuan Aquaculture Brigade performed a productive test. 50 kg of pig exereta (pigsty effluent occupies 50%) or 50 kg of nightsoil can produce 56 kg of terrestrial grasses. 50 kg of net pig manure can produce 112 kg of grasses.

The Food Conversion Coefficient of English rye grass and Sudan grass is 18.4–27.5, average 27. Thus, 100 kg of pig excreta through growing grass can be converted into 4.2 kg of herbivorous fish, bring along rearing 1.5 kg of filter-feeding and omnivorous fish, the total yield is 5.7 kg. According to the tests of Freshwater Fisheries Research Centre, 100 kg of pig excreta can be converted into 2.5 kg of fish, whereas 50 kg of net pig excreta through growing grass can produce 3.2 kg of fish, increasing 133 %. In pig-grass-fish integration, the dominant species are herbivorous fish, which can meet the demand of the market and gain much profit. 100 kg of pig excreta are used to grow grass then to culture fish, resulting in the increment of 2.90 Yuan which is 1.34 times that of fish farming by applying pig excreta directly.

Allocation of Pig-grass-fish

In the combination of pig-grass link, the guide line is to facilitate grass planting and manure application. If grasses are planted on pond dykes only, the simple pig sty can be built near the pond; if there is a fodder field beside pond dyke, centralised pig shed near fodder field is taken into consideration. Firstly, the ratio of pig and grass depends on unit-area yield of pasture grasses. Below the critical level of high yield, the more the yield, the more the number of pigs. Secondly, the ratio also depends on the utilisation of pond silt. The more pond silt is used, the fewer the number of pigs. Thirdly, it depends upon the quality of soil and pig manure and varieties of pasture grasses. On the basis of production test for ryegrass and sudan grass grown all the year round, the best allocation is 5–6 pigs per mu. The combination of fish-grass link has already been introduced in fish-cum-crop integration.

Web of Integrated Management

In order to raise the utilisation efficiency of energy, to decrease cost, to increase products and income, several models could be interwoven to an integration web through longitudinal and transverse nutrients cycles and energy flow in line with the local conditions. There are three kinds of integration web so far.

Multi-level Comprehensive Utilisation Web

In the light of different utilisation of nutrients by various specialised trader the trades can be connected in series with nutrient cycle and energy flow as a master line. An integration wed can be regarded as a number of connected nutrient cycles and energy flow.

That is to say, in recycling of chicken manure, the spilt and undigested chicken feeds can be fully utilised. Sgan Bei Livestock and Aquaculture Farm adopts this kind of integration web. Not only can chicken manure be fully utilised but also it can save about 25kg of fine feeds for pig, increasing net income by 47 yuan/pig. Furthermore, through simple deodourisation and sterilisation, chicken manure is mixed with fine feeds as composite pellets, which can be used to feed chicken again and then chicken manure can be fed to pig.

In Zhejiang Province, there is a common multi-level comprehensive utilisation model, that is, cow-mushroom-earthworm-duck-fish or cow-mushroombiogas-earthworm-duck-fish. Cow dung can be used to grow mushroom; cow urine for fish farming; mushroom soil for earthworm culture; and then, earthworm is fed to ducks, the excreta of which is applied to fish ponds. The levels can be rearranged in line with the local conditions. Thus, different constituents of cow excreta are utilised at different levels or the new constituent produced at the preceding level can be utilised by the next level.

Parallel-cennected Web of Multi-trades

With fish farming as the centre, there may be a parallel connection of multi-trades, such as fish-livestock-crop integration to create sources of feeds and fertilizers. The products and wastes are utilised reciprocally among all trades. Most of integrated fish farms adopt this kind of integration web. E.g. in Xiang Yang Aquaculture Farm, fish farming is combined with pig, cow and geese raising.

Animal manures which is self-supporting are used for fish farming and for crops such as pasture grasses, squash, sweet potatoes and soybean and at the same time for cultivation of aquatic plants such as water hyacinth, Spirodella spp. and Wolffia arrihza which are used as feedstuff for fish, pig and geese.

Although soybean is planted on pond dyke in the latter part of a year, water hyacinth can offer 80% of green fodder for fish and pig.

Fish-livestock-crop and Agriculture-industry-commerce Web

On the basis of parallel connection of fish-livestock-crop integration, all trades develop on both input and output ends. On input end, feedstuff processing industries can be set up and food can be utilised at multilevels; on the output end, the products of all trades needs processing and marketing.

The waste of processing factories is used to raise livestock and poultry. This is a combination, which can raise the level of comprehensive utilisation of natural resources and the rate of utilisation of energy and can increase the products of fish, livestock and poultry resulting in the increment of value, income and job opportunity.

In 1970's, this type of integration web was set up in Dong Hu Fish Farm, Siang Yin County, Hunan Province. In recent years, the management structure and the scale of this web have developed, further, e.g. Helei Fish Farm has developed into an aquaculture, industry and commerce combination.

On the axis of abscissae, they integrated fish farming with grass cultivation, pig, cow and duck.

On the axis of ordinates, they process crops, pupae so as to provide green fodder, pupae waste water for fish, livestock and poultry at the end of input; at the end of output, they process the products of livestock. e.g. making preserved egg, slaughtering ducks. The waste of processing factories can be used as feeds for fish, livestock and poultry or fertilizers for fish ponds. The processed products go to market home and abroad. Thus, the economic efficiency can be raised greatly.

For example, in 1982, the average net profit was 6.78 Yuan/duck. The people who are engaged in duck raising and egg processing occupy 17% of the personnel of the farm while the net profit accounted for 43%.

Animal Raisig and Plant Cultivationon an Integrated Fish Farm

Animal Raisingon an Integrated Fish Farm

The Purpose of raising animalson an integrated fish farm is to develop integrated fish farming and fully utilise the limited feedstuffs. The multi-stage utilisation of feedstuffs and fertilizers makes it possible to supply the community with more produce and to increase the income for the fish farm as well. Such activity in China is practised in line with local conditions, that is to say, different natural resources and farm conditions decide different items of management and animal raising, e.g. various integration of fish - livestock - poultry or integrated management of "aquaculture, agriculture, or a composite of animal husbandry, industry and commerce."

Sericulture

Pond dykes, corner plots, etc. can be used to plant mulberry tree for sericulture on an integrated fish farm. The by-products of sericulture, dregs, which are comprised of the faeces and sloughs of silkworm and mulberry residues can be used as fertilizers for fish culture. On average, we can get 50 – 60 kg silkworm dregs out of 100 kg mulberry leaves after feeding silkworms. If they are used to feed fish, 1,000 kg of silkworm dregs can be converted into 100 kg of fish. By applying 5,000 kg of pond silt to mulberry plots, we can increase 250 kg of mulberry leaves, which can, in turn, raise the production of silkworm cocoon. Pupa occupies 80% of a wilkworm cocoon by weight, and 100 kg of raw silk and 600 kg of pupae can be obtained from every 700 – 800 kg of silkworm cocoon. Silkworm chrysalis is very rich in protein and fat, which are the good feedstuff for fish, livestock and poultry. Pupae have a special ordor, the feeding amount should be controlled. The nutritive components in silkworm drege and pupae are as follows (in percentage):

The collection and utilisation of by-products from sericulture:

Silkworm dregs, mulberry residues could be collected at an early stage of silkworm breeding, but the amount is negligible. It is the right time to collect them after the dormant stage. Silkworm pupae are byproducts after reeling. There are two ways to collect above-mentioned by-products:

1. Sundry method: It is suitable for sunny days. Pulverise them after dry, and mix them with chicken feeds for broilers. The daily amount should not exceed 5% of the total feeds. This must be discontinued when the chicken are one month old, otherwise, the meat quality would be affected. The amount of pupae should not be over 10% of pigs' compound fodder. The pupae are allowed to feed sows, piglings, not hogs.
2. Water immersion method: It is for rainy days. As it is difficult to dry silkworm dregs and pupae, just stock these in big vats and fill the vats with water. The preserved ones can be fed to piglings only, never for the fatteners because of the stronger odor of the wet pupae, which may influence the quality of pork.

Chicken Farming

The digestive tract of a chicken is very short, only 6 times of its own body length. Some of their feedstuffs are excreted out before being fully digested.

From the research results, it is found that about 80% of the feedstuffs are utilised and digested by the poultry (count by dry material); Therefore, various animals n the integrated fish farm can use 20% undigested feedstuffs from the chicken manure. Besides, chicken has the habit of pecking foods, more than 10–15 % of the feed are scattered over the ground these can also be reclaimed together with the chicken manure. The total protein contents of the dry chicken is as high as 20–30%. In it is satisfactory if the protein contents of feedstuffs are over 18%. In the past chicken manure used to be fertilizer for the crops which, we consider, is a world of waste.

A little slip will cost a big loss. It is very different from small scale family raising.

The utilisation of good species and hybrid vigor of broilers that are suitable for an integrated fish farm: The most common species of broiler chicken in the world are hybrid offspring of White Cornish and White Rock. Some companies in the world are specifically engaged in chicken breeding. They have got some purebred for the hybrid experiments. From the results of their hybrid matching, they select the best combination of the hybrids. They have first generation hybrids (grandfather generation), and second generation hybrids (parent generation), and then produce broiler chickens for commercial production.

The white feather broilers bred in China are Starbro System from Canada and Hybro System from Holland. We have established National Pure line Chicken Farm, selling grandfather generation to the first grade chicken farms, and the first grade chicken farms sell parent generations to the second grade chicken farms. The second grade chicken farms supplies double hybrid commercial chickens for the markets.

Chinese people prefer to eat yellow feather broiler chickens, therefore, most of the white feather broiler chickens are processed as fronen chicken for export.

The live poultry for domestic market and , yellow feather chickens. They are "New Pudong Chicken" bred by the Institute of Rusbandary, Shanghai Academy of Agriculture. The meat of this species is better than Starbro chicken imported, but their growth is a little slower, their body weight reaches 1,500 grams after 10 weeks, a hen can lay 140 eggs per year. Their viability is.

In recent years, some of the developed countries use processed chicken manure to feed animals and poultry. "Chicken Manure Pelletining Production" which becomes the "Regenerative technology Industry" of 80's is of very important economical value. In China, dry saw-dust and pulverised dry stalk of crops are used as bedding material for chicken coops.

The bedding material and chicken excreta are used to feed fish. Part of them may be eaten directly by fish as feedstuffs while the rest will become the nutrients of the fish pond water. In this way, we still cannot fully utilise all the nutrients of chicken manure. For a change, the excreta together with the bedding material are used as the feedstuffs for cattle or pigs, especially for sows and beef cattle.

Thus, the undigested and unabsorbed fine feeds in chicken excreta are well utilised again by sow and cow, promoting the development of the hog and cow husbandry. The excreta of pigs and cattle can also be used to feed fishes.

Fishes can utilise the undigested nutrients in pig manure and cow dung. The utilisation rate of feedstuffs is then greatly increased.

Besides, animal manures serve as fertilizers in the water to produce abundant plankton, which is a good natural food for fish. If animal excreta are used to culture fish, they enrich the pond silt, which can be used as fertilizer for the mulberry plot or on the pond dyke for green fodders resulting in high yield. Thus, we can raise silkworm, also raising chicken, pig, cow and rearing fish at the same time. In this way, the utilisation of the feedstuffs and fertilizers can reach its maximum, and aquaculture and animal raising can be carried out in a more comprehensive way.

Rearing of Fry (From Fry to Summerlings)

Application of Base Manure

After pond clearing and before the stocking of fry, base manure should be applied to the pond so as to propagate the planktonic community (such as rotifera, nauplius and micro-zooplankton, etc.) as proper feeds for fry. After manuring, various zooplanktons proliferate fast in the pond. The peak orders are as follows: protozoa—rotifera and nauplius—micro-cladocera—macro-cladocera-copepoda. Therefore, fry should be timely stocked in ponds so that they can get enough palatable natural food at each growing stage. The applying time and quantity of base manure are dependent on the type of manure. If the decayed manures or compost are used, they are applied 3—5 days before stocking with the amount of 150—300kg/mu. And if Dacao or green manure are used, they are applied with the amount of 200 — 300kg/mu, 5—10 days before stocking.

After stocking, additional manure should be applied so as to stabilise the fertility of the pond water. At the initial stage (fry are only reared for the first 10 days with a body length of less than 2 cm), fry should be ensured with enough natural food of rotifera and micro-cladocera and their biomass are 20-40mg/l. At the late stage (Fry have been reared for more than 10 days and their body length reaches more than 2 cm.) while applying manure to propagate zooplankton, the biomass of phytoplankton required for culturing Silver carp is 30 mg/l and the commercial food is supplemented for Grass carp and Bighead at the late stage.

Stocking Density

The growth and survival rates of fry are closely related to stocking density. That is to say, they are affected by food, water space and water quality. The optimum stocking density for Silver carp and Bighead fry is 100,000—120,000 ind/mu, Grass carp and Black carp fry 80,000—100,000 ind/mu. The lower stocking rate of fry will waste water body and increase the production cost. The stocking density can be appropriately increased with skilful farming techniques and careful management.

Pond Fertilization

Significance of Fertilization in Pond

China has a long history of pond fertilization for fish culture ages ago. The farmers adopted the methods of manuring to rear fry. For instance, Dacao (green manure) has been applied in Guangdong and Guangxi Provinces and night soil has been used as manures in Jiangxi and Human Provinces so as to nurture fry into summer fingerlings. Rich experience has been accumulated. In fingerling rearing ponds and table-fish rearing ponds, fertilization is aimed at developing natural food and saving artificial feeds. This method has been widely adopted in China. Phytoplankton, elementary producers in ponds take in all essential inorganic materials from the water and conduct phytosynthesis with sunlight, producing organic nourishment needed for their own growth and reproduction.

Fertilization means to supply phytoplankton with nutrients for phytosynthesis and to promote the growth of phytoplankton, by which zooplankton and other aquatic animals are fed on for their growth and propagation. Fish feed on plankton and other hydrobios. Pond fertilization lies in cultivation of various food organisms and their propagation in large quantities in fish ponds to provide fish with abundant natural feeds, by which they can grow faster. The yield of fish pond may be raised thereby.

Among the fish food organisms in ponds, there are a series of interrelations between the predator and the prey. The biological term is “food chain”. Fish is the last link of the food chain of hydrobios in ponds. Usually, animals could use only 5–20% energy of both animal and plant feeds. Utilisation of energy is related to the length of food chain. The shorter the food chain is, the higher the rate of energy transfer will be, in other words, the higher the utilisation rate of energy transfer is, the higher the fish production will be. The hydrobios in ponds are in a constant progress of growth and death.

The dead bodies of organisms will turn from complex organic materials into simple inorganic materials through decomposition by bacteria, then dissolve in water, which can again be utilised by phytoplankton to produce new individuals. Hence, the materials in ponds are in a constant state of circulation mainly through the food relationship between living organisms. This kind of circulation is called pond material circulation.

The process of pond material circulation is just that of the production of fish and their food organisms in ponds. The circulation originates from the soluble minerals in water. Under certain conditions of light and heat, the propagation of phytoplankton depends upon the amount of nutrients in water. In turn, the production of all food organisms depends upon the proliferation of phytoplankton. Fertilization is just to enhance the quantity of nutrients and to nurture food organisms in ponds so as to create favourable conditions for raising fish yield.

Kinds of Organic Manures and Methods of Application

Organic manures mainly refer to excrements of farm animals. It is a general term of manures containing organic matter. Nowadays, manures applied in fish ponds in China are mainly organic manures. In production, the following are often used, faeces and urine of livestock and poultry, night soil, green manure, compost and silkworm faeces etc.. Only through decomposition by microorganisms, organic manures may be converted to nutrients, which plants can absorb, and then, organic manures are full of nutritional elements with long effects.

Bibliography

Atre, P. K.: *Fish Genetics and Aquatic Environment*, Navyug, Delhi, 2008.

Bailey, MT: *A-Z of Tropical Fish Diseases and Health Problems*, Howell Book House, New York, 2001.

Bennett, M. V. L.: *The Central Nervous System and Fish Behavior*, University of Chicago Press, Chicago, 1968.

Breder, C. M., and Rosen, D. E.: *Modes of Reproduction in Fishes, How Fishes Breed*, Natural History Press, New York, 1966.

Burgess, Peter: *Tropical Fishlopaedia: A Complete Guide to Tropical Fish Care*, Howell Books, NY, 2000.

Chandra, Prabhas: *Fishery Conservation, Management and Development*, SBS, Delhi, 2007.

Deka, Manab: *Fish Fermentation: Traditional to Modern Approaches*, New India Pub, Delhi, 2009.

Devi, Kamla: *Poisonous and Venomous Fishes of Andaman Islands, Bay of Bengal,* Zoological Survey of India, 2003.

Dholakia, A.D.: *Fisheries and Aquatic Resources of India*, Daya, Delhi, 2004.

Emygdio, L. Cadima: *Fish Stock Assessment Manual*, Daya, Delhi, 2007.

Farrand, John: *The Audubon Society Master Guide to Birding,* Alfred A. Knopf, New York. 1983

Frank, C. Edminster: *Fish Ponds for the Farm*, Agrobios, Delhi, 2004.

Geran, James: *The Proper Care of Goldfish*, TFH Publications, NJ, 2000.

Grizzle J.M.: *Anatomy and Histology of the Channel Catfish*, Auburn Printing Co, 1976.

Hall, G.M.: *Fish Processing Technology*, Springer, Delhi, 2009.

Hargrove, Mic, David Brown, and Maddy Hargrove: *The Discus: An Owner's Guide to Happy Healthy Fish*, Howell Books, NY, 1999.

Hasler, A. D.: *Orientation and Fish Migration, in Hoar, W. S., and Randall, D. J.,* New York, Academic Press, 1971.

Heda, Nilesh: *Fresh Water Fishes of Central India: A Field Guide,* Vigyan Prasar, 2009.

Helfman G., Collette B., & Facey D.: *The Diversity of Fishes*, Blackwell Publishing, UK, 2004.

Ingles, Lloyd G.: *Mammals of the Pacific States,* Stanford University Press. 1965

Kulkarni, G.K.: *Fisheries and Fish Toxicology*, APH, Delhi, 2006.

Laureti, Edmonodo: *Fish and Fishery Products: World Apparent Consumption Statistics Based on Food Balance Sheets (1961-1995),* Daya for FAO, Delhi, 2001.

Lewbart G.A. *Self-Assesment Color Review of Ornamental Fish*, Iowa State Press,1998.

Marshall, N. B.: *The Life of Fishes*, Weidenfield and Nicholson, London, 1965.

Nelson, J. S.: *Fishes of the World*, John Wiley & Sons, Inc.. , NY, 2006.

Pollard, D. A. *Freshwater Fishes of South-eastern Australia*. McDowell, R. M. Reed, Sydney. 1980.

Poppe T.T., *A Color Atlas of Salmonid Diseases*, Academic Press, London, 1996.

Ralph, R.: *Breeding Australia's desert goby*, Tropical Fish Hobbyist, 1988.

Roberts, R.J: *Fish Pathology*, Bailliere Tindall, London, 1989.

Robertson, D.R.: *Fishes of the Tropical Eastern Pacific,* University of Hawaii Press, Honolulu. 1994.

Schmida, Gunther E.: *Rainbow Fish*. Barron's Educational Series, NY, 2000.

Selvamani, B.R. and R.K. Mahadevan: *Fish and Fishery Culture: Assessment and Evaluation*, Campus Books International, Delhi, 2008.

Sharma, L.L. and A.K. Mathur: *Hand Book of Freshwater Ornamental Fishes,* Yash Pub house, Delhi, 2006.

Silva, De: *Fish Nutrition in Aquaculture*, Springer, Delhi, 2001.

Srivastava, C. B. L.: *Fish Biology*, Narendra, Delhi, 2008.

Srivastava, M.P.: *Natural History of Fishes and Systematics of Freshwater Fishes of India,* Narendra Publishing House, Delhi, 2002.

Trevor, A. Anderson: *Fish Nutrition in Aquaculture*, Springer, Delhi, 2009.

Untergasser D.: *Handbook of Fish Diseases*, T.F.H. Publications, Neptune, NJ, 1989.

Ward, Barbara E.: *Varieties of the Conscious Model: Fishermen of South China*, Tavistock, London, 1965.

Index

❑❑❑